Study Skills

Study Skills: A Student Survival Guide

The Institute of Cancer Research

Edited by

Kathryn L. Allen
Director, Interactive Education Unit
The Institute of Cancer Research, Sutton, UK

Contributors

Jeff Bamber
Maggie Flower
Barry Jenkins
Bob Ott
Stan Venitt
Neil Walford
Steve Webb

John Wiley & Sons, Ltd

The Institute
of Cancer Research

Other Wiley Editorial Offices

John Wiley & Sons Inc., 111 River Street, Hoboken, NJ 07030, USA

Jossey-Bass, 989 Market Street, San Francisco, CA 94103-1741, USA

Wiley-VCH Verlag GmbH, Boschstr. 12, D-69469 Weinheim, Germany

John Wiley & Sons Australia Ltd, 33 Park Road, Milton, Queensland 4064, Australia

John Wiley & Sons (Asia) Pte Ltd, 2 Clementi Loop #02-01, Jin Xing Distripark, Singapore 129809

John Wiley & Sons Canada Ltd, 22 Worcester Road, Etobicoke, Ontario, Canada M9W 1L1

Wiley also publishes its books in a variety of electronic formats. Some content that appears
in print may not be available in electronic books.

Library of Congress Cataloguing-in-Publication Data

Study skills: a student survival guide / editor, Kathryn L. Allen.
      p. cm.
   ISBN-13 978-0-470-09485-3
   ISBN-10 0-470-09485-0
1. Science–Miscellanea. 2. Doctor of philosophy degree–Handbooks, manuals, etc. 3. Study skills–
Handbooks, manuals, etc. 4. Universities and colleges–Graduate work–Handbooks, manuals, etc.
5. Time management. I. Allen, Kathryn L.
Q173.S78 2005
502–dc22   2004027089

British Library Cataloguing-in-Publication Data

A catalogue record for this book is available from the British Library

   ISBN-13 978-0-470-09485-3 (PB)
   ISBN-10 0-470-09485-0 (PB)

Typeset in 10/13 pt Bembo by TechBooks, New Delhi, India

Contents

Contents

Contents

Preface

This book started life as a website for PhD students at The Institute of Cancer Research, the content of which has been adapted and expanded for a wider general science student audience.

A number of people have helped make the journey from web to print a smooth one. I would like to thank all the students and scientists at The Institute who have been involved in the development of the book. Special thanks go to Laure Parent, Jane Torr, Corey Gutierrez, Charlotte Westbury, Rachael Natrajan and Ian Collins for reviewing the chapters and providing helpful feedback and suggestions.

Thanks also to Liz Goad for all her help with developing the book and for our many useful discussions on the content and format. I am grateful to Sue Sugden, who played an important role in developing the material for Chapter 3 (Information Retrieval).

Last but not least I would like to thank Jane Sewell, for her research work and input into the structure of the book, as well as her meticulous proofing and checking of the manuscript and Catherine Dunbar for all her help with the checking process.

Kathryn Allen

Foreword

Most of you have probably recently completed an undergraduate degree during which you were constantly swamped with reading lists. And now here you are as a new PhD student, eager to get on (we hope) with cutting-edge scientific research. But what is the first thing you are given? More reading.

Why then should you spend more time reading about how to do a PhD before you actually start one? Although doing a PhD can be thrilling – particularly for those of you who have never done original research before – it can also be a difficult and lonely time. Before long those dreams of scientific fame (perhaps fortune too) can seem impossibly remote. Your supervisors and colleagues will certainly give you all sorts of good advice, but they can't anticipate all the various pitfalls that you may encounter. You can't train for a PhD; you can only do it. This book, however, is the next best thing. You can think of it as your survival guide. But it's more than that. This book also provides the tools that you will need to get the most out of your PhD so that you can fulfil your ambitions. Want to win a Nobel Prize? This is where you start!

The book has been arranged so that you don't have to read it all in one go. I suggest that you skim the whole thing now and read Sections One and Two carefully. These are the sections that you will need to get started. Then get stuck into your research and have some fun. But don't neglect this book; return to it periodically. When you do, ask yourself a few questions:

Am I being efficient?

Am I interacting well with my research team?

Have I been able to access all the information I need?

And, above all, am I getting results?

If you answer 'yes' to all these questions, then you are doing well, and before long you will be asked to present your work and write it up for publication. That's the time to read Section Three in more detail. It will make these tasks less daunting and improve your presentations and papers. And then, just when you think that you have the whole thing sorted, you will come to the final hurdle of a PhD: writing the thesis. This just shouldn't be painful. Read the last chapter (Section Three) of the book, and you will find that it is really rather easy and even fun. Promise.

A PhD is not a matter of just doing science; it is about learning how to do science. There is a difference, and this book tells you what it is. When you understand that, you will enjoy your PhD more, be more productive and provide yourself with the basis with which to build a research career. I wish you the best of luck.

Clare Isacke
Head, Molecular Cell Biology Team
Breakthrough Breast Cancer Research Centre
The Institute of Cancer Research

Introduction

This book is a quick reference study skills guide aimed at PhD science students. It identifies the skills you need to progress successfully through your PhD and on into your working life. The book presents practical advice and tips, in an easy-to-use format designed for students with busy schedules.

The featured skills are those that complement your technical knowledge. They are important because they enable you to organise, utilise and communicate your knowledge effectively. They may take some time to learn but utilising these skills will have an impact on your efficiency, your quality of life and stress levels, and your future.

How you learn to organise your work or the way you feel most comfortable making presentations, for example, is individual to you. You will learn from practice and experience. The advice provided in this book is best used as a guide, to help you to find the way you work best and exploit your abilities to the full.

Each of the chapters has been written by authors from The Institute of Cancer Research, all of whom have either supervised students working towards their PhD or are involved in providing services for them. Additionally, all the chapters have been reviewed by students currently undertaking a PhD at The Institute and their useful suggestions have been integrated into the finished product.

The book is split into three main sections, Personal Effectiveness, Finding and Using Information, and Communication Skills. Each of the sections is made up of chapters containing practical advice and quick reference tips, as well as specific points to remember. Don't feel you have to read the sections/chapters in any particular order; all the skills will be useful to you throughout your PhD project. Yes, it is a good idea to organise your work and your time from the start of your studies, but don't leave thinking about writing your thesis until the last minute – remember that your thesis is the end product of a sustained effort over all your years of work.

SECTION ONE: PERSONAL EFFECTIVENESS comprises chapters on how to organise your work and your time and on developing your interpersonal skills – your behaviour and how it impacts on others. All the skills and advice assembled in this section are focused on you and how you work.

Good organisation is fundamental to your studies; planning, not lack of ability, is often where students fall down. Co-operative working relationships are also important, both for support and advice and this demands that you are aware of yourself and how your behaviour affects others. You can learn skills to help you foster productive relationships in your immediate working environment and develop a network of contacts within the wider research community.

SECTION TWO: FINDING AND USING INFORMATION focuses on the need for you to identify, locate and read a lot of literature and having a strategy for each of these activities will make your task easier. The chapter on information retrieval is an overview of where and how to look for information and also provides some practical tips on evaluating your information sources, as well as managing the large number of references you are likely to accumulate. Once you have identified which papers are important to your work you need to critically review their contents. Critical reading – an essential skill for eliciting and evaluating information from relevant primary research papers – is covered in the second chapter of this section.

The third and final section, **SECTION THREE: COMMUNICATION SKILLS**, looks at how you communicate your work, orally, visually and in writing. The chapters in this section cover some of the ways and settings in which you will present your work and the skills you need for the job. The aim of the section is to provide pragmatic advice on some of the most demanding aspects of your PhD – writing and defending your thesis, writing papers, and making oral and poster presentations. These are skills you will continue to use throughout your scientific career.

At the end of each chapter there is a checklist and, where appropriate, a summary of references, further reading and web resources relevant to the chapter subject. Additionally, throughout the book you will find references to websites provided as further information sources. The nature of the Internet is such that websites and their addresses (URLs) change frequently. The URLs provided in the book were correct at time of publication, although some change is likely to occur over time.

We hope that you find the guide easy to use and of practical value during and after your studies. If you have feedback on any aspect of the book then please e-mail your comments to *ieu@icr.ac.uk* or call 00 44 20 8722 4370.

Kathryn Allen

List of contributors

Dr Jeff Bamber, Head of the Ultrasound and Optical Imaging Physics Research Team, The Joint Department of Physics, The Institute of Cancer Research and the Royal Marsden NHS Foundation Trust

Dr Maggie Flower, Senior Lecturer in Physics as Applied to Medicine, The Joint Department of Physics, The Institute of Cancer Research and the Royal Marsden NHS Foundation Trust

Mr Barry Jenkins, Librarian, The Institute of Cancer Research

Professor Bob Ott, Head of Radioisotope Physics, The Joint Department of Physics, The Institute of Cancer Research and the Royal Marsden NHS Foundation Trust

Mr Neil Walford, Training and Organisational Development Manager, Human Resources Department, The Institute of Cancer Research

Professor Steve Webb, Head of Radiotherapy Physics Research, The Joint Department of Physics, The Institute of Cancer Research and the Royal Marsden NHS Foundation Trust

Dr Stan Venitt, Emeritus Reader in Cancer Studies, University of London

SECTION ONE: PERSONAL EFFECTIVENESS

Chapter 1
Effective Organisation and Time Management

by Professor Steve Webb
Head of Radiotherapy Physics Research, The Joint Department of Physics, The Institute of Cancer Research and the Royal Marsden NHS Foundation Trust
and Professor Bob Ott
Head of Radioisotope Physics, The Joint Department of Physics, The Institute of Cancer Research and the Royal Marsden NHS Foundation Trust

Introduction

At the start of your PhD, you may feel like you have all the time in the world, but it's amazing how quickly the years can pass. Before you know it you will be sitting in front of your computer trying desperately to remember how you performed that experiment 18 months ago, and what the result was. So, from the outset, your main aim should be to manage your work and your time as effectively as possible.

This chapter provides practical advice on organising your work and the successful management of your time while you work through your PhD. You will find some tips on reducing stress and maintaining a balance between the demands of work, social and family life, which is important and often overlooked.

Before you embark on the chapter, take some time to consider how you currently work. Doing a simple activity like the one below can help you to identify ways in which you may be working inefficiently – most people have habits and routines that prevent them working as effectively as they could.

ACTIVITY

Keep a diary of your daily activities for a week and see how you are ACTUALLY using your time. And, try to identify any personal traits that you think prevent you working effectively, for example, you might be:

- a procrastinator – putting off tasks and letting them pile up
- distracted – by interruptions or other demands on your time
- disorganised – surrounded by mountains of papers or flitting from one task to another

---

- a perfectionist – missing deadlines to perfect work
- too optimistic – taking on too much or setting unrealistic deadlines
- impatient – needing to do everything immediately even if it means working excessive hours.

All of these problems can be avoided with good planning and organisation, coupled with the ongoing advice of your supervisor.

Organising your work

Good organisation is fundamental to your PhD studies. Going hand-in-hand with effective time management, organisational skills help you to get the most out of your time.

Keeping track of events in a work diary, knowing where you've put all those references and how you've named your computer files will give you more control over your day-to-day activities. You will also feel the benefit in the longer term, especially when you come to write your thesis.

Your institution may suggest a portfolio approach to organising your work and time. This can be used to set out the objectives of the project, which are planned and agreed with your supervisor, and then details any further developments and activities as a sort of extended diary of events. This will help you and your supervisor to follow the progress of your project and highlight any missing elements. Much of the practical advice in this chapter is applicable if you are producing a portfolio.

Maintaining a good lab book

A lab book is your course notebook and will become an invaluable resource to you during your PhD. As you carry out experiments, it is vital that each is precisely recorded to help trouble-shoot any problems as you go along. It is also important to date your lab book, make note of any observations and to keep track of batch numbers of any materials you use. Writing everything down in a lab book not only gives you a permanent record of your experiments and results, but will also act as a prompt when you come to write your thesis.

Maintaining a written record is also important if, for example, you make a ground-breaking discovery. If this occurs you will need to be able to prove that you carried out the experiment and provide exact details of the methods and equipment you used.

Although it may seem like a chore, writing up as you go along (experiment by experiment or day by day) will actually benefit you in the longer term as you won't have to take time trying to remember what you did last week/month/ year.

Don't worry if your book isn't neat — it is not meant to be a masterpiece. The book is a permanent record of your research, and you should write down any thoughts, ideas or plans there and then, before they slip your mind. Avoid using 'post-it'^{TM} notes or scrap paper as these can easily be lost.

Tips for keeping a good lab book

- Always keep lab books for as long as you are continuing your work:
  - at least until you have successfully defended your thesis
  - your institution may keep them for longer
- Date everything:
  - this is useful if you are relating a particular experiment to another piece of work
- Keep lab books in a suitable form, so they can be shown and explained to someone else:
  - this is particularly important for collaborative work
- Stick in graphs, gel photos, images:
  - loose items can fall out and be lost
- Use one book per project or sub-project where appropriate:
  - mixing projects can be confusing
- It is good practice to write a summary of the progress of the project every now and then:
  - particularly after a long period of day-to-day recording
  - if you are producing a portfolio, these records are useful for project reports and reference material
- If you write a computer program to make some calculation, document changes that are made and ensure that any output stays linked timewise to the appropriate source code. If you do not do this then it will be hard to relate data that you generate to the source which generated it. (This problem did not arise in the old days when line printers were used and the source code program and its data stayed physically attached.) It is also a good idea to make a note of which variables in lab book mathematics correspond to which computer variables.
- Don't feel disappointed if you don't appear to have many results:
  - they often come in fits and starts and can be cumulative — sometimes progress only becomes apparent on review of the information in your lab book
- Extract results from your lab book into small reports before meeting with your supervisor:
  - give your supervisor a copy a day or two ahead of the meeting
  - after the meeting, document what was discussed and action to be taken, and note it in your book

Keeping your computer files in order

The computer files you generate during your PhD will grow in number and size beyond your expectations. It is therefore important to establish directories and subdirectories at the beginning of your work, and add new ones as and when they are needed.

It saves time if you can easily locate the file you need. This will be particularly important when it comes to writing your thesis.

Valuable work (e.g. your thesis/reports/research papers) should ALWAYS be backed up. Back up your work onto a floppy disc, zip disc, CD or your network drive area, or all of the above! Such files are precious, and you can never be too careful – make sure you do this regularly.

E-mail also needs organising – keep folders of e-mails relating to the same person or topic. An e-mail can be used as written evidence of a communication, since it is now more common to e-mail than send a letter.

Tips for maintaining organised computer files

- Use subdirectories:
  - keep one project separate from another
- Use a name for each program/file that gives some idea of what it does or contains
- Review your files and e-mails regularly and delete anything that is no longer useful:
  - this will reduce the number of files and e-mails you have to trawl through to find the one you're looking for
- **ALWAYS** back-up your files, on floppy or zip discs, or CD:
  - never keep all your files on one computer ('eggs in one basket')
  - keep a copy of files somewhere physically away from your department, just in case there's a fire or robbery that could prevent you from accessing the computer that stores your work
- **ALWAYS** put your name in the program and the date it was started; record updates in the program (with dates) and relate these back to notes in a logbook
- Keep the main program short and use as many subroutines as possible
- If the program looks nothing like the hardcopy maths, provide a table of symbols that can be related to the program maths:
  - provide lots of comments
- Keep all source code:
  - deleting it could be a hindrance in future work

- Avoid 'knock-on errors':
    - don't get into a situation where, for example, you fix one subroutine but forget it is linked into several main programs that haven't been linked yet

Handling reference papers

You will be overwhelmed with reading materials throughout your PhD. You need an efficient filing system so you can locate papers quickly when you need them, and keep track of where your papers are if you lend them to people.

Reference Manager and Endnote are useful computer databases that allow you to document reference papers and recall them according to topic or author name (see 'Managing your references' in Chapter 3, page 52). You can also write notes to yourself against each reference, so you can record where you found the reference, the main points of the paper or any other details.

Tips for managing reference papers

- Invent a practical system for filing that you understand:
    - alphabetically by author – easy to find a paper but tricky if you want to locate a subset of papers by topic
    - filed by project – good for locating subsets related to a project, but tricky to find by author
- Choose a method of storing your papers that suits you:
    - filing cabinets – can store lots of material, but are not easily transferable
    - buff folders or box files – these store fewer papers, but are easier to transport, e.g. to take home
    - mixture of both – get the best of both worlds, but be careful not to become confused and lose track of your papers
- Keep your own copies of papers:
    - annotate them and highlight important facts/results
    - keep a note of the overall essence/take home message of the paper, for example as computer files or index cards. This is useful for quick reference to papers – attaching the printout or index card to the front of the paper can help you to remember key points
- Start a lending list:
    - if people borrow your papers or books it's quite possible you may never see them again

- keep a list of where the paper is and who has it, also the date you lent it out
- Make a note of what you're reading in your lab book:
    - this promotes a feeling that your reading is working and aiding your research
    - it records when you read something that may be important
- Keep up-to-date with publications:
    - look at a subset of journals regularly, either by doing database searches or by signing up for current awareness services (see 'Current awareness services' in Chapter 3, page 49)
    - follow up on papers that others suggest are relevant to your studies

Why keep a diary?

As a student you may feel you don't have many appointments to record in a diary. However, a diary can be used as a back up to your lab book as well as to record meetings and appointments. You can use it to plan ahead for the week, and at the end, compare your plans with what you've actually achieved.

Tips for keeping a good diary

- Write down all your appointments:
    - so you can keep track of what activities are due or planned and also what you have done
- Record all deadlines:
    - this will help you to manage your time and plan ahead for deadlines
    - you can also write notes such as 'see five days ahead – report must be ready by then', to jog your memory
- Make a note of any periods of time your supervisor will be away:
    - this becomes particularly important in your final year – there is no point giving your supervisor a chapter of your thesis to read if he or she is going on holiday in three days time
- Divide each day into two parts:
    - you can write major appointments or reminders in one part and use the other part for a two to three-line summary of what you plan to do/have done that day
- Write a summary at major calendar landmarks:
    - for example, Christmas, Easter, before you go on holiday
    - write a 'what I've done since last summary' short account – it helps you to keep up-to-date with your progress; you could plan to do this at the same time as writing your lab book summary

- Have a 'rolling items' list:
  - use a separate notebook or a card for your list
  - the list is intended for items that need to be done but for no particular deadline
  - make sure your list evolves and that you cross things off once you've done them

Organising your time

The key to optimising your use of time is **planning ahead** – from day to day, to year on year – for the duration of your PhD.

An overview of the planning phases

Each stage of your PhD requires planning so that you can allocate sufficient time to each task throughout your years of study, and provide yourself with short-term and long-term aims. You need to think about what you want to achieve on a daily, weekly and yearly basis, for example:

Daily and weekly planning

- Read
- Familiarise yourself with the theory of an experiment; plan and carry out experiments (being aware of specific equipment/materials you may need to book or order)
- Prepare presentations e.g. for lab meetings
- Attend any courses or lectures
- In your first year, ensure that you develop the appropriate skills required to carry out your project

Monthly planning

- Read and prepare the materials to be discussed at your tutorials
- Plan which courses to attend and keep a record of those you have attended or plan to attend; keep up-to-date with any course notes
- Plan which seminars to attend or prepare your materials if you are to present one

Yearly planning

- Plan for annual reports; keeping up-to-date with work throughout the year makes report writing much easier and stress-free

- Have an idea of how your thesis will be written. In your final year, have a firm plan for completion of the chapters of your thesis, to ensure writing-up is not a last-minute rush
- If you are writing or contributing to a research paper, have clear ideas of the milestones to be met in its preparation
- Plan for conferences to attend during the year, preparing poster/oral pre-sentations if necessary and allowing sufficient time to do so
- One way to make sure you keep up-to-date with planning and carrying out the project is to produce a portfolio as mentioned earlier in this chapter

Daily planning

Use a daily plan to record what your daily tasks are, how important they are, and how long you feel they should take. These plans are a useful way of taking stock of what you need to do. They can also help to determine the effectiveness of your time usage in relation to the importance of the task.

Tips on planning your day

As you are devising your daily plan consider the following:

- How and when you work best should affect how you plan your daily tasks. Are you an early riser who falls asleep in the evening, or are you normally half-asleep in the morning but can work into the evenings? The key is to work when you are most alert, and try to plan your day around your optimum hours
- A lot can be achieved early in the morning, in the evenings and at week-ends, when there is less pressure on computers and equipment. Be pre-pared to work the unpopular hours if this gets your work done ahead of time
- Before you start, make sure you have all the information you need. Find a place to work where you are least likely to be disturbed, such as your own lab, office or at home. Also, where possible, try to plan around interruptions
- Find out when your supervisor will be available to provide guidance or answer questions. If you know the best times to see your supervisor, you won't waste time trying to locate him or her
- Additionally, your hours should fit in with:
    - co-workers
    - equipment availability
    - seminars/tutorials/lunchtime lectures
    - reading time

• Try to make sure you include breaks in your schedule for relaxation and contact with other people. Too much out-of-hours working is bad for morale. Strike a balance, work hard but ensure you have time to play and, importantly, to sleep

 Remember, if you are working out-of-hours in the laboratory, it is important that someone knows that you will be there alone as there are health and fire-safety issues to be considered. Make sure that you understand the working-alone rules, inform your supervisor when this happens and fill in the appropriate forms if necessary to allow you to do so.

Planning experiments

When planning an experiment, it can be helpful to ask yourself some simple questions to focus your thoughts. The answers to these questions will have a bearing on what you do and when you do it. Such questions include:

• What do I want to achieve this week?
   • how will I divide up the stages of the experiment between the days?
• How many hours do I want to put in?
   • do I have any social engagements that will prevent me working an extra hour or two tonight?
• How do the hours break down into different tasks?
   • for example, how long will it take me to set up, run and get the results of this experiment or how long does it take for the computer program to produce results?
• How important is each of the tasks?
   • prioritise so that you put your efforts where they are most beneficial to the project
• Am I doing anything unnecessary?
   • for example, is there a task that can be done by someone else or left out completely?

Be flexible

You might not be able to stick rigorously to your work plan; sometimes things don't run smoothly, such as the availability of equipment or materials, computing requirements or an experiment takes longer than you anticipated. Where possible, try to allow for such factors when planning your work.

Overall

• It is important to plan your work but don't spend too much time doing it, to the detriment of the work itself

- Make sure that the planned experiments fit the priorities and objectives laid out in your overall plan, meetings with your supervisor or portfolio (if you have one)
- Set realistic deadlines for specific pieces of work – if in doubt, discuss with your supervisor or someone else who has carried out the same kind of work
- Always allow for contingencies, such as equipment breakdown or factors outside your control

You may find it useful, especially in the early stages of your PhD, to make use of a plan that highlights and prioritises your tasks. Figure 1a shows a tabular daily plan with tasks time-limited and in order of importance.

Project planning

There are several stages involved in planning a project, and time needs to be allocated to each of them.

Tips on project planning

- Read in preparation:
  - try to read around a project thoroughly before you start any practical laboratory work – this will save time in the long run, and lessen the chance of having to repeat an experiment because of unforeseen problems
  - it is important to read and write up continuously throughout your PhD – this will help to overcome a last-minute rush in the final three to six months
- Define project objectives:
  - before you start any practical work, devise some objectives and aims for your project. It helps to be clear from the outset how you want the project to progress. Setting objectives should be covered in your meetings with your supervisor
  - as the project evolves some of your objectives may change, or new ones may arise – again, use your meetings with your supervisor to review these
  - some objectives may also be dropped if they become less relevant, or are less useful than others – talk to your supervisor before making any major changes
- Determine facilities needed:
  - precious time can be wasted searching for equipment or setting up an experiment and then finding that the equipment you needed is already in use – plan what equipment you will be needing at what stage of the experiment and book it

FIGURE 1a Example of a daily plan

Page	Day	Date	Month	Year
1	2	22	4	2005

Start Time	Finish Time	Activity	Task Rate	Mins per task A	B	C
9.30	10.30	Read paper to clarify methods for next experiment; book PCR machine	A	60		
10.30	11.00	15 mins discussion with staff member re: ongoing project	B		15	
		15min calculating oligo concentration for next PCR experiment	A	15		
11.00	12.15	Preparing material for tutorial tomorrow	A	75		
12.15	1.00	Lunch	A	45		
1.00	2.00	Setting up PCR, making reaction mixes, run machine, make gel	A	60		
2.00	3.30	Library – find and photocopy papers for reading for tutorial next week	B		90	
3.30	3.45	Coffee break	A	15		
3.45	4.15	Load and run agarose gel with PCR reactions	B		30	
4.15	5.15	Read papers from library and make notes	C			60
5.15	5.30	Take photo of gel and label with reaction names	A	15		
		Hours		4	2	1
		Minutes		45	15	0
		DAY TOTAL	8 hours			

KEY

A – high-priority task

B – medium-priority task

C – low-priority task

- you may need to buy some materials in from outside suppliers, or wait for things to be made for you, for example. The delay between ordering and delivery is beyond your control and should be taken into account when planning an experiment
- Plan the mix of experiment, writing and theory:
  - make sure the balance between these is sufficient to allow optimal time management; you don't want to waste unnecessary time reading when you could be setting up an experiment and then reading about the next stage while it is in progress

Long-term planning

Every year of your PhD will involve deadlines of some kind, related to your annual reports, posters for conferences or contribution to research papers. It is important, therefore, to allow time for incidental deadlines during your PhD, as well as the known deadlines such as reports, tutorials and meetings.

It's often useful to consider the three or four-year duration of your PhD in sections, such as three-month periods, and to allocate an aim for each of the 'Quarters'. For example, in Quarter One you should aim to:

- do a fair amount of reading for your project
- acclimatise and familiarise yourself with staff/facilities
- attend courses that will be helpful to your PhD
- get yourself set up on your institution's computer and network system (e.g. e-mail account and personal folders).

Keep writing . . .

Continue writing throughout your PhD – for example, tutorial notes and essays can become incorporated into your final thesis. Certain sections of your thesis, such as the introduction and materials and methods, can often be written quite early on. You can always tweak them later if necessary. If your supervisor asks you to write a review of the field you are studying in the first months of the study, this could easily form part of the background chapter to your thesis. Your supervisor will then have a chance to assess your scientific writing skills at an early stage, allowing any problems to be rectified.

A general project plan

An idealised three-year PhD plan is shown in Figure 1b. You may not be able to adhere rigidly to a plan like this. The major events shown in bold will be applicable

FIGURE 1b Generalised project plan		
Background reading and planning	October–April	Yr 1
Coursework	October–April	Yr 1
Preliminary work	January–September	Yr 1
Skills and technique learning	Before first year report	Yr 1
First year report	September	Yr 1
MPhil to PhD transfer process/ end of first year probation	October–January	Yr 2
Second year report	September	Yr 2
Major laboratory/ research work	October (Yr 2)–December (Yr 3)	Yr 2 & 3
Thesis plan	End of December	Yr 3
Writing-up	Start early February	Yr 3
Submission to university	End of September	Yr 3
Viva	0–3 months post submission	
Corrections	Allow 2 weeks to complete	

to all PhD students, although the timing might be slightly different, depending when you began your studies and whether you have a four-year grant.

Wherever possible, you should aim to reach the milestones in the time suggested to avoid a last-minute rush during your final year – try to write-up as you go along.

This plan is an overview of the milestones in your PhD – they may not apply to everyone, and many are flexible. For instance, it is essential to read about your field throughout the project to ensure that you are up-to-date with the latest developments. It is possible that your PhD might be scooped by someone else or even be proved a waste of time in the worst case, if you ignore what others are doing (see 'Current awareness services' in Chapter 3, page 49).

There is no magic formula for being successful in your PhD studies, but the best way is to keep doing a bit of everything throughout the three or four years – a bit like doing revision regularly rather than leaving it all to the end, but on a larger scale.

 Remember, during the course of your writing-up you may need to repeat the odd experiment here and there, but avoid extensive lab work in the last few months when you are in the final stages of writing up.

Stress and the work/life balance

Establishing a good work/life balance is an important part of being a student. Finding the time for work, friends and family and your own interests is important. Each of these aspects should not necessarily get the same amount of time but a good balance of work, play and sleep is needed to make sure that stress is minimised. Continuously working long hours is not the best solution to most problems – it makes you tired and reduces your productivity, often with the result that your work is of a lower standard.

You will probably encounter stress at some point during your project – things do not always go well with the work, people seem to get in the way, things happen outside your work with family and friends. If you experience stress, take action to combat it as early as possible; if you don't, it could seriously detract from the success of your project and cause health problems later.

Tips for avoiding and dealing with stress

- Practise good time management and project planning:
  - preparation takes a lot of the uncertainty out of your project
- Take regular exercise:
  - try relaxation exercises as well as challenging workouts; daily exercise is ideal
- Talk to someone about being stressed out, so that it doesn't become unmanageable
- Eat a balanced diet:
  - avoid quick-fix sugar and stimulants like sweets and caffeine
- Maintain a healthy balance of work, play and sleep
- Be positive:
  - give yourself a pep talk every now and again (see 'Using self-talk to develop assertiveness' in Chapter 2, page 22)

Checklist

✓ Keep good lab books, paper and computer files:
  * keeping organised files will help when you come to write your thesis
✓ Use meetings with your supervisor to help set aims and objectives, and determine whether you have achieved them
✓ Use a diary to keep appointments, record deadlines and plan your work
✓ Write up as you go along, wherever possible
✓ Read around your project at the start and throughout your PhD
✓ Finish a piece of work/experiment and write it up straight away:
  * don't be caught out in two years time with unfinished work
  * materials and methods sections can be added to whenever you try a new experiment
✓ Don't spend endless time planning
✓ IF IN DOUBT, ASK – make the most of others' experience:
  * tutor/supervisor
  * post-docs in your lab
  * technicians or any others who may be able to help
  * other students who might overlap with your project or are using similar techniques
✓ Maintain a good work/life balance
✓ Combat stress early

Finding out more

Books

Cottrell, S. (2003) *The Study Skills Handbook*, 2nd edition, Palgrave Macmillan, Basingstoke.

Websites

Taylor, A., Turner, J. and Collier, J., *The University of Reading study guides: Time management – organising yourself and time.* Available at:
http://www.rdg.ac.uk/counselling/studyskills/publish/study%20guides/time.htm

Chapter 2
Personal and Interpersonal Skills

by Mr Neil Walford

Training and Organisational Development Manager, Human Resources Department, The Institute of Cancer Research

Introduction

Good personal and interpersonal skills will be invaluable to you throughout your student years and on into your career. They enable you to foster productive and co-operative working relationships and build the network of contacts necessary for your success.

Exactly how important these skills are is often underestimated. Evidence shows that employers are looking for specific 'soft' skills to complement 'hard' or technical knowledge, seeing them as a package. Your ability to, for example:

- assert yourself and your ideas
- work as part of a team
- network
- show initiative

is seen as an indicator of your future effectiveness at work. And employers have developed techniques to test candidates' soft skills in a number of ways beyond the conventional interview, using psychometric testing for example.

Personal and interpersonal skills come under the umbrella of 'emotional intelligence', defined as the personal characteristics, skills and competencies that are responsible for the ways in which you behave, how you feel, how you relate to others and how well you perform in your job.

How you decide to work on your personal and interpersonal skills is individual to you. What you will find in this chapter are the basic tenets of the skills, how they can be developed and some tips on putting them into practice.

Self-awareness

To be able to develop yourself, you need to be aware of your current strengths and weaknesses. There are many exercises you can do to assess these, all variations on the same theme. Undertaking an assessment activity like the one below will help you identify skills areas you may need to develop. The featured list contains a selection of the key personal and interpersonal skills and knowledge areas that have been identified by the Research Councils and the Arts and Humanities Research Board as necessary for a successful research career.

ACTIVITY

Here is the list of skills under different subject headings (excerpted from the *'Skills Training Requirements for Research Students: A joint statement by the Research Councils and the Arts and Humanities Research Board')*:

Personal effectiveness: a willingness to learn and acquire knowledge:

• Creative, innovative and original in approach to research
• Flexible and open-minded
• Self aware, ability to identify own training needs
• Self-disciplined, motivated and thorough
• Recognises boundaries and uses appropriate sources of support
• Self reliant, can work independently

Communication:

• Clear appropriate writing style e.g. progress reports, articles, thesis
• Constructs coherent arguments and articulates ideas clearly to range of audiences (formally and informally using a variety of techniques)
• Defends research constructively to internal and external audiences
• Contributes to wider public understanding of research field
• Effectively supports learning of others e.g. teaching, mentoring or demonstrating

Networking and team working:

• Develops and maintains co-operative working relationships with manager, colleagues, peers and wider research community
• Understands own impact and contribution to formal and informal teams
• Gives and receives feedback constructively

Explore your level of skill against each of the items on the list by rating yourself on a scale of 1 to 4:

1 = A personal strength – you could/do assist others in this area
2 = Fully competent – you are able to meet the requirements of your role
3 = Some development areas – in some aspects you need more experience or knowledge
4 = Major development area – you have a lack of experience/knowledge.

Once you have done the assessment, try giving the results to a peer or colleague for additional feedback. Even if you don't agree with the feedback you receive, you will gain an insight into the way others view you – this can be invaluable information, particularly if you spend time working in teams.

Assertiveness

Assertiveness is a vital skill. As a student, you need to learn to communicate your thoughts and needs clearly so that you can manage your life and work successfully.

Assertiveness is the clear communication of your thoughts, feelings and needs – often collectively referred to as your rights – balanced with the recognition of the rights of others. Being confident enough to express yourself in this way is an invaluable tool.

Remember that you have the right to:

- **say no**
- **be the final judge of your own thoughts, behaviour and emotions**
- **have and express an opinion**
- **make mistakes**
- **negotiate**
- **put yourself first sometimes**
- **stand up for yourself, or not, if that's what you choose.**

Practise being assertive; it is an everyday skill and there are many instances when it will be useful to you throughout your PhD, for example, in meetings with your supervisor, working in a team, settling publication authorship or defending your work/thesis. In these situations, it is important that your needs or ideas are clearly articulated and understood. But assertiveness is not just a question of pushing your ideas or needs forward at all costs, it is a skill that requires tact and an appreciation of individual circumstances.

Assertiveness should not be confused with aggressive behaviour. When you are aggressive you are domineering, often angry and insensitive to others. Assertiveness means clearly stating your needs . . .

- at a time when it is most appropriate
- in a way that is appropriate
- having regard for the rights of others; a balancing of mutual rights and responsibilities.

There are going to be times when you may choose to be aggressive or submissive (as distinct from assertive) according to circumstances; the balance between all three and when they are appropriate, needs to be clear in your mind.

By learning to be assertive and practising these skills, you will build relationships that are honest and productive, giving you a greater degree of control over your working life – while reducing unnecessary stress.

> ## Tips on assertiveness
>
> - Use direct and honest language – say what you mean
> - Talk in the first person and use simple statements
> - Don't let others impose their ideas or views on you but respect the feelings and ideas of others
> - Be tactful
> - Be aware of your body language and that of others, and try to maintain eye contact
> - Give others an opportunity to air their views, and be open to feedback
> - Learn to say no – where this is appropriate
> - Try to deal with compliments or criticism as pleasing or useful information

Using self-talk to develop assertiveness

As any competitive athlete will tell you, the benefits of positive thinking cannot be underestimated. Self-talk is a way of promoting positive thinking. It is the internal conversation that you have with yourself, an internal dialogue that triggers emotions and then actions. Self-talk can be either negative or positive and 'talking yourself up' or 'down' will have an effect on the outcome of your interaction with others and on your own performance.

ACTIVITY

Below is an example of how self-talk can work. When you have read it, think about a situation in which you have behaved either negatively or positively. Try to break the situation down into its component parts so you can see both how the situation evolved and how your thoughts may have influenced the outcome.

On receiving unfair negative criticism from your superior you say to yourself that this is unfair, you feel injustice and your behaviour reflects that. As you respond passively, becoming defensive and avoiding eye contact, the outcome is that the person delivering the feedback feels justified in that criticism.

Turn that situation on its head; the same unfair criticism is delivered but this time you are prepared to receive it. You say to yourself that you understand it is not personal and you need to find out why your supervisor is criticising you and know how to respond. Talking to yourself in this way makes you feel in control, focusing on what your manager is saying and on whether there has been a misinterpretation or misunderstanding. Your behaviour is assertive; you are listening, trying to understand and acknowledge the

---

concerns, and helping your supervisor to understand the facts. The situation can be satisfactorily resolved and ground rules have been laid for future interaction.

Self-talk is not only a way to develop assertiveness but also a great way of motivating yourself and also helping to control nerves – before presentations for example. When using self-talk in this way, repeat a few short positive statements to yourself to boost your confidence, for example, '*I am confident*', '*I am capable*'...

Effective listening

Listening well is at the root of interpersonal skills; feedback, teaching, and team working all depend on it, to varying degrees. Listening improves your knowledge and appreciation of a subject, situation or person, but it also improves accuracy, minimising mistakes and wasted time. Although you may think you listen automatically and without effort, it is a skill and requires thought and practice.

Different circumstances influence the way you listen, requiring different approaches. You may be attending a conference, a workshop or a meeting and this will determine the way you listen.

For instance, at conferences or seminars you may need to make very little – or a passive – contribution, whereas in other events there are active exchanges – meetings or workshops perhaps. Even on a one-to-one basis, both active and passive listening may be required.

Tips to help you listen (and remember) more effectively

One-to-one	Larger groups
• Face the speaker	• Face the speaker
• Try to maintain eye contact	• Stay focused
• Try to relax while staying alert	• Keep an open mind – put aside your own bias or prejudice
• Keep an open mind – put aside your own bias or prejudice	• Don't interrupt – wait for a natural pause before asking questions
• Don't interrupt – wait for a natural pause before asking questions	• Take note of the speaker's body language
• Give regular indications that you are listening	• Avoid distractions like background noise and even your own thoughts
• Take notes to remind you of the main points	• When listening for long periods, concentrate on the key issues

- Wait and listen to the whole meaning before you speak

- Take notes to help you remember what was said, or plan to tell someone about what you've heard
- Listen critically – evaluate the information and its delivery

 Remember to:

- **wait and listen for the whole meaning before you speak**
- **listen for the main points**
- **avoid distractions**
- **stay focused**
- **look for what is not spoken.**

Giving and receiving feedback

Feedback is a way of providing information, choices and advice. As a student you will experience the feedback process in both formal and informal settings, for example as part of your supervision, during presentations at conferences, while teaching or when working as part of a team. Whatever the setting, feedback is a necessary and highly productive way of developing yourself and your work, but it is important to give and receive feedback in the right way and in context.

Giving feedback

Here are some golden rules to follow when giving feedback:

- before starting, **always** consider the value of what you're saying to the person receiving the feedback – be motivated by the desire to help
- give personal feedback face to face – never via e-mail for example
- offer feedback on what **you** know or have seen
- don't be judgmental – give descriptions of what you actually saw or felt, and use 'I' statements
- choose the points that are most important and limit yourself to these
- ask questions rather than make statements
- set ground rules and inform the person of the subject matter of the session in advance – don't spring surprises
- feedback must be timely i.e. close in time to the event to which it relates
- highlight areas where the person did well, as well as areas for improvement
- if you have negative feedback to give, try to place it between two pieces of positive feedback
- be concise and specific – don't waffle

- observe everyone's personal limits
- ask for feedback at the end of the session to give the person the opportunity to respond.

Try to avoid:

- providing incomplete or second-hand information
- exaggerating
- placing blame
- springing surprises.

Receiving feedback

There are plenty of occasions where you will find yourself on the receiving end in the feedback process, particularly as a student. It may take the form of constructive criticism of your work from your supervisor, defending your thesis and maybe feedback from students you are teaching. Be open to criticism and also to compliments; here are some points for you to consider:

- don't be defensive even if you feel it – allow others to be honest
- stay focused
- listen
- ask questions to help clarify comments
- acknowledge valid points
- take time to think about what you hear
- take notes
- return to the person giving the feedback with suggestions or ideas about specific points raised.

Relationships and team working

Relationships will be important to you during your studies and later when it comes to looking for work. These relationships may be with your peers and colleagues, your supervisors, or your professional or network contacts. You need to pay attention to these relationships and maintain and develop them to ensure that:

- you are supported and have both formal and informal forums for discussion of your work
- your professional needs are met
- your reputation and opportunities are maximised.

Peers and colleagues

The research and lab environments provide a network of people with whom to interact. You may rely on them for support and advice or as valuable team members and working partners. It is important that you develop co-operative relationships with those you work and interact with, and recognise that you are always working as a team, whether formally or informally.

Try to get to know a mixture of people from both inside and outside of the lab. Post-docs, research assistants and other PhD students can prove invaluable sources of knowledge and experience, as can your peers and colleagues. It is always useful to know who to approach for advice or information on an informal level.

More formally, you will find yourself working in a team, which may be made up of colleagues, peers and possibly your supervisor (see 'Your relationship with your supervisor' later in this chapter). When you are working as part of a team, it is useful to understand something of its dynamics and practical workings before you start. For the team and its members to benefit from team-working, some basic preparation or training is advisable prior to undertaking the group activity. Here are some starting points to consider . . .

Tips for working as part of a team

- Make sure that each of you are clear about the role of each person in the group, as well as your own
- Agree some ground rules about decision-making and problem resolution
- Respect others in the group and their opinions:
  - value and support different views and new ideas
- Understand group dynamics:
  - take time to look at the situations and relationships within the group (see 'Teaching small groups' later in this chapter)
- Actively participate and make a contribution:
  - give and receive feedback
  - listen
  - ask questions as a way of encouraging new ideas
- Involve others in your planning and decision making
- Be prepared to 'muck in' to get the job done
- Resolve conflict
- Respect the group's decision

Getting the most out of team meetings

Like them or not, meetings are a fundamental part of team communication, discussion and decision-making.

The role of the person chairing the meeting is to encourage a contribution from all participants, ask questions to clarify, summarise regularly, keep the meeting to time, to manage conflicts of opinion within the group and to make sure that actions are correctly recorded and followed up.

Tips on running and contributing to meetings

Running a meeting	Contributing to a meeting
• Only the relevant people should attend • The goal of the meeting should be clear, as should what is expected of the participants • There must be an agenda clearly stating who will lead discussion on each item and the start and finish time • All relevant information should be sent to the participants in advance • When the meeting is long, try to have regular breaks • Appoint a chairperson to manage and moderate discussion • Structure the meeting; have an introduction, then aim to set out the issues, debate and then summarise • Try to manage disruption, acknowledge people's feelings without allowing these feelings to take over the meeting • Minutes should be written up and circulated as soon as possible after the meeting	• Give a clear introduction to your points, let people know what you're going to say so they know what they're going to hear • Using visual aids will help any presentation • Always substantiate your points • Try to identify the concerns and experience of the other participants so that you can reflect or address these when you speak • Use influencing skills where you can; acknowledging the views of others or supporting other speakers verbally or non-verbally encourages support in return

Your relationship with your supervisor

Your supervisor is your main reference point during your PhD, and you may also have a back-up/second supervisor to review your work regularly. The relationship with your supervisor is an important and a personal one, with no hard and fast rules governing how it should work. How the relationship works is very much dependent upon the individuals concerned. However, there are some basic, practical points to consider so that interactions with your supervisor are both harmonious and beneficial.

Ideally, your supervisor should be accessible, but is unlikely to be freely available so try to find out in advance the times when he or she is most likely to be free as this will save you time in the long run. He or she needs to be aware of your project and its progress, helping to give it direction and being willing to discuss ideas, references and resources. Enthusiasm and support of your work and career are important but you should also receive constructive criticism of your work and its presentation, with your supervisor inputting his or her knowledge of science in general and your specific field. Through your supervisor you should be aware of the standards required for your work, and you should meet other workers in your subject area and learn of useful conferences. Your supervisor needs to read your thesis prior to submission and tell you when you are ready to submit.

Also remember that supervisors should suggest opportunities for collaborative work. Such opportunities could include Co-operative Awards in Science and Engineering (CASE) studentships – which are collaborative research opportunities between academic and commercial partners – or working for a lab that already has links with a commercial entity. Your supervisor should also advise you on issues surrounding the uptake of that work, such as intellectual property rights for example. For more information about the CASE see *http://www.epsrc.ac.uk*.

If things go wrong

Students' relationships with their supervisors are usually trouble-free. But if the relationship between you and your supervisor does break down, the most important thing is to deal with it as early and tactfully as possible. First, talk it through together. If that proves unproductive, approach your second supervisor. Consider all the possible solutions and options, as changing supervisor is not easy. However, if you cannot resolve the difficulties, your institution will have a formal procedure for you to follow. Take a look at The Institute of Cancer Research's procedure as an example:

> 'The Head of Department is responsible for ensuring the provision of good supervision and facilities for the entirety of the project. The Head of Department will be the first point of call in the case of any difficulty relating to the resources for or supervision of a project. Where possible, the Head of Department will remedy the difficulty. If this is not possible, the matter will be referred to the Deputy Dean who will decide upon and oversee the implementation of a solution which may include, in extraordinary circumstances, a change in supervision or location.'

Tips for managing your relationship with your supervisor

- Find out in advance when your supervisor will be available and try to arrange a meeting once a week or once a fortnight. Plan time to see your supervisor into your work schedule
- Make the most of these meetings:
    - prepare ahead
    - tell you supervisor in advance what you want to discuss and take your list to the meeting with you, as a reminder
    - talk through any difficulties (do this as near as possible to when they arise)
    - set objectives – assigning and reviewing deadlines
    - set times and dates for your next meeting
    - be direct and communicate clearly
    - make notes
- Make sure you're aware of any departmental or university procedures that apply to you as a PhD student
- Be enthusiastic about your work
- Take full advantage of the resources and facilities available to you
- Take your supervisor's advice unless you can justify acting otherwise
- If you are producing a portfolio, this will give you a basic framework for your relationship with your supervisor and for developing a work plan (for more information on portfolios see 'Organising your work' in Chapter 1, page 4)

Networking

The University of London Careers Service (*www.ucl.ac.ukl/careers*) identified networking as one of the major sources of employment for research scientists with more post-docs obtaining their next job through their networks (33%) than through advertisements (25%)[1].

Networking is a crucial skill in the research world as it is a close-knit community that places great emphasis on peer-review and makes substantial use of personal references for recruitment and promotion.

Obvious networking opportunities such as conferences are a major part of academic life, giving you an opportunity to update your professional knowledge as well as publicise your work, identify future collaborators and raise your profile among your peers. You need to ensure that when the right opportunities present themselves, your name and reputation are known to the relevant people.

---

[1] UCL Careers Service (2003) *'First steps to the future'*

Tips on networking

- Always seek to present your work:
  - using seminars or poster presentations for example
- Read the list of delegates before attending conferences:
  - identify those you want to speak to
- Introduce yourself:
  - don't stay with those you know, get used to approaching people; what are the key points you want people to know about you?
- Be prepared to talk about your work:
  - think about how you would describe your work and your skills
- Follow up on those who show an interest in your work
- Take business cards with you to events
- Once you have made contacts, maintain them:
  - keep a record of your contacts
- Be open to opportunities for collaboration

Remember, constantly develop your network and maintain it.

Teaching small groups

Many students teach during their PhD, either as a requirement of their course or for financial reasons. Whatever the reason you decide to teach, it is a great opportunity to develop your skills base. For you it is an exercise in communication, leadership and feedback – skills increasingly required in the workplace. For your students, it is a forum for discussion in which they can develop their ideas, improve their understanding and work together.

Before you start, get it in writing . . .

There are a few things that you need to be clear about before you start. Always make sure that the details of the job are clear and in writing, i.e. the hours you will be required to work, what is expected of you and how much you will be paid. You need to know this information so that you can plan ahead, working the teaching into your schedule, without over-burdening yourself. Once you are clear about what is required you can start.

Teaching skills

For you and your students to get the best out of the experience you need to communicate clearly within the framework of a teaching style and have some concept of how groups function. Some of the information on presenting in Chapter 5 will also be relevant to talking to small groups.

Teaching/learning style

Your main aim is to get your point across while keeping your students interested. When you are preparing for a session, choose a particular teaching style to provide your group with the structure they need to work within. Try to use a variety of teaching styles or methods over time to prevent your students from getting bored. There are many methods you can employ; rotate them to motivate and encourage group participation/interest.

You can use a tutorial approach for small groups (to approach a particular problem the students have been working on), problem-solving or discussion groups, role play (giving students an opportunity to argue from a particular standpoint), facilitated group work, student-led groups, brainstorming or presentations by groups or individuals.

Tips on teaching

Once you have decided on a style for your session, try to think about how you want it to progress. This will utilise your organisation and people/communication skills:

- Identify clear outcomes for each tutorial:
    - when setting direction and required outcomes for the group, make sure you plan ahead and ask for feedback from others, including your students
- Ensure the tasks you set are manageable and achievable in the time you have
- Ensure your students or audience know what is required of them
- Prepare supporting information
- Involve your audience/students
- Ask questions:
    - not just 'why' or 'what', but different types of questions to encourage different types of responses
- Listen carefully
- Provide appropriate feedback and provide it skilfully:
    - make sure you foster and encourage ideas
- Work towards participation from all:
    - encourage reluctant students while limiting the contribution of very verbal students
- Be prepared to provide an explanation of complex points

- Be aware of how teams work:
  - it pays to familiarise yourself with factors that can govern group behaviour (see 'Group dynamics' below)
- Look for non-verbal signals – body language, lack of group cohesion

Group dynamics – how the group works

As well as the process and content of the session, it is useful to be aware of the ways that group members interrelate, i.e. how the group works. Group dynamics can affect the way group members communicate and work together. If you pay attention to the dynamics of the group, you give it the opportunity to maximise its potential.

Some basic points to consider are:

- group size – the size of the group, the mix of its members and their expectations are some initial factors that may affect how the group members interrelate and the level of their individual participation
- seating arrangement – the location and layout of the meeting and seating can present a formal or informal setting for the group. It is particularly important to pay attention to seating, from the point of view of communication and participation, make sure the seating arrangements are conducive to open discussion
- ground rules – structure and procedure contribute to the smooth running of the group. Setting aims and tasks within a framework of ground rules and clear procedures allows any conflict or disagreement experienced within the group to be managed effectively.

Checklist

✓ Learn to be assertive when you need to be
✓ Understand your strengths and weaknesses to help guide your personal development
✓ Learn to listen and give/receive feedback effectively
✓ Build up your internal and external network
✓ Ensure that you are well supported in your teaching
✓ Practise your personal and interpersonal skills

Finding out more

Articles

Hunter, J. E. and Hunter, R. F. (1984) Validity and utility of alternative predictors of job performance, *Psychological Bulletin*, **76 (1)**, pp 72–93.
Kelly, R. E. and Caplan, J. (1993) How Bells labs create star performers. *Harvard Business Review*, July-August.

Books

Belbin, R. M. (2003) *Management Teams – Why they succeed or fail, 2nd edition*, Butterworth Heinemann, Oxford.

Forsyth, P. (1996) *Making Meetings Work*, Institute of Personnel and Development, London.

Gillen, T. (1997) *Assertiveness*, Institute of Personnel and Development, London.

Sternberg, R. (1997) *Successful Intelligence*, Plume, New York.

Goleman, D. (1995) *Emotional Intelligence: Why it can matter more than IQ*, Bantam Books, New York.

Goleman, D. (2001) *Harvard Business Review on what makes a leader*, Harvard Business School Press, Cambridge, Massachusetts.

Honey, P. (1988) *Improve Your People Skills*, Institute of Personnel Management, London.

Hopson, B. and Scally, M. (1991) *Build Your Own Rainbow: A life-skills workbook for career and life management (Lifeskills for adults)*, Mercury Business Books, London.

Lewis, G. (1997) *Institute of Management Open Learning Programme: Developing yourself and your staff*, Pergamon Flexible Learning, Oxford.

Nierenberg, A. R. (2002) *Nonstop Networking: How to improve your life, luck and career*, Capital Books, Sterling, Virginia.

Phillips, E. M. and Pugh, D. S. (2003) *How to Get a PhD. A handbook for students and their supervisors, 3rd edition*, Open University Press, Maidenhead.

Stewart, I. and Joines, V. (1994) *TA Today: A new introduction to transactional analysis*, Lifespace Publishing, Nottingham.

Whetton, D. A. and Cameron, K. (1991) *Developing Management Skills, 2nd edition*, Harper Collins, New York.

Websites

Jaques, D. *Small group teaching*. Oxford Brooks University, Oxford Centre for Staff and Learning Development. Available at:

http://www.brookes.ac.uk/services/ocsd/2

learntch/small-group/sgtindex.html

UK GRAD programme website. Available at: http://www.grad.ac.uk

(Courses and workshops for UK post graduate students, focusing on group working and interaction.)

University of Glasgow Teaching and Learning Service website. Small group processes I & II. Available at: http://www.gla.ac.uk/services/tls/NLP/Small/

SECTION TWO: FINDING AND USING INFORMATION

Chapter 3
Information Retrieval

by Mr Barry Jenkins
Librarian, The Institute of Cancer Research

Introduction

One important part of achieving your PhD is that you be aware of and consider the existing literature in your subject area, enabling you to present a rationale and context for your own work. You will also need to keep abreast of the literature throughout the course of your studies.

The wealth of information available means that you need to identify and evaluate sources effectively, as well as search efficiently. No single source can provide everything you need to know, so you must use and search a range of resources to ensure you don't miss information. Learn to make the most of the resources and search facilities offered by your institution's library.

Your library may offer training on how to research databases and locate information from various online and printed sources. The following guidance broadly covers some key information sources, but not all institution libraries will subscribe to the same sources; so check your library website for up-to-date information, for example, changes to web addresses.

You will also need to compile a comprehensive bibliography for your thesis, citing the literature consulted during your research project. It makes sense to collate and manage your references as you go along, and some practical guidance on this is also provided here.

Information sources

There are many different resources you may draw on for information during the course of your studies; these are detailed below.

Primary and secondary sources

As a PhD student your primary information sources will be those containing peer-reviewed primary research. Primary research could be a set of experiments that have tested a stated hypothesis, or the results of a clinical trial. Most papers produced from this work will be published in peer-reviewed academic journals like *Nature, Cell, Physical Review* or journals of the American Chemical Society. Peer-review means that articles have been evaluated by an editorial panel and scrutinised by independent experts in the particular field. An article that has been through this process has been 'quality assured' (see 'Peer-review' in Chapter 4, page 55).

Some primary research may be published in textbooks or conference proceedings as full papers or abstracts, but it is unlikely that the work will have been peer-reviewed in this format and these publications are not considered primary sources.

A secondary source collates, discusses and reviews primary research on a specific topic and often comes in the form of textbooks, conference proceedings or review articles. Secondary sources can be a useful way to get the background or an overview of a subject area; a review article, for example, reviews all relevant papers recently published on a topic and is published in a journal. Remember that you still need to go to the papers referenced in the bibliography to read the full details of the original research.

Textbooks and e-books

Textbooks are a valuable information source. They are useful for background information, providing a broad overview and introduction to a subject. When choosing textbooks, look for:

- reference books and encyclopaedias – many are now available in e-format
- standard work which has run to many editions
- work providing an introduction and overview of a subject by an author or editor from a prestigious institution
- books receiving favourable reviews in good scientific journals.

Works that are not widely published and have not been subject to peer-review, while still useful, are not reliable enough to use as primary reference sources, these include:

- conference proceedings – these may be brief summaries of research, which could be published more fully in later research journals
- collections of papers from a journal – these are fine for an overall survey of the topic but are usually not fully edited
- works from minor publishing houses – see the websites listed below for the major scientific publishers.

Some publishers now produce online versions of selected books. Your institution library will need to subscribe to these but they do mirror the printed versions and have the advantage of being accessible at all times for all readers, unlike popular print books which may be on loan from libraries.

Some of the major scientific publishers' websites include:

- Cambridge University Press – *http://uk.cambridge.org*
- Elsevier – *http://www.elsevier.com*
- Humana Press – *http://www.humanapress.com*
- Lippincott, Williams and Wilkins – *http://www.lww.com*
- Oxford University Press – *http://www.oup.co.uk*
- Wiley – *http://www3.interscience.wiley.com*

You can search these sites by subject or title and most will display the publication details and an abstract of the book. In addition, your library website may produce core-book lists for particular subjects.

The British Library (*http://catalogue.bl.uk*) holds the best catalogue if you are trying to locate specific textbooks. For the equivalent in the US, see the catalogue of the Library of Congress (*http://www.loc.gov*). The major UK university libraries have their catalogues freely available on the web. For a full list and links, see the Library & Information Services section on the National Information Services and Systems website (*www.hero.ac.uk/niss*).

Journals and e-journals

The traditional published journal remains the primary information source for peer-reviewed research. Most journals are now available online and are a good way to get the full text of an article. You can find papers via the contents pages. Using your library network you can access both the journals to which your library subscribes and those that are freely available.

Your library may have 'IP controlled' access to journals, in which case no password is required when accessing the full-text versions within your institution. Where access is password controlled, your library may use ATHENS, which provides a sign on and has the advantage of allowing remote access, i.e. from computers outside the institution. The major full-text database for journals in the sciences is ScienceDirect (*http://www.sciencedirect.com*), encompassing the thousands of journals published by Elsevier.

 Remember, subject searching for e-journals in one database like ScienceDirect is a good starting point but will not provide you with comprehensive listings of all journals. ScienceDirect, for example, only

features those journals published by Elsevier; you may miss relevant information if you limit your subject search to a particular publisher in this way.

Although most journals are available in electronic format, back issues of journals more than 20 years old are unlikely to be accessible electronically. It is important that you consult the classic key references in your subject area in the original hard copy of the journal. Although most libraries don't stock back issues, nearly all journals will be available through your inter-library loan service.

Open access journals

An open access journal is an e-journal that makes its full text content freely available. Researchers, particularly in the US, are lobbying publishers to make their content free for all to access on the web.

The authors of the papers in open access journals pay for them to be published and some institutions take out subscriptions to the providers so their researchers can publish without being charged. These journals are all peer-reviewed and have the obvious advantage of making research available to everyone as soon as it is published.

The following two publishers are currently the major open access providers:

• BiomedCentral (BMC) – *http://www.biomedcentral.com*
• Public Library of Science – *http://www.plos.org*

Most of the BMC journals are indexed in PubMed, the US National Library of Medicine database, which will link you directly to the relevant websites when performing a search (see 'Bibliographic databases' later in this chapter).

In the physical sciences, the arXiv.com e-print archive (*http://xxx.lanl.gov* or *http://lanl.arxiv.org*) publishes full text papers of current research before they appear in the traditional journal format. It covers the fields of physics, maths, computer science and quantitative biology.

In addition, some major journals are either freely available (e.g. *British Medical Journal*) or provide free access to papers that have been published for longer than a certain period of time, e.g. 6, 12 or 18 months. At present, many publishers only have electronic full-text versions for the last five years or so. However, publishers are starting to put their full archive online, for example, the *Journal of Biological Chemistry* has its full archive online, dating back to 1905 – it is freely available on the web, except for recent years that are still subscription based.

For a comprehensive list of the main scientific journals and their free full-text content, see the 'Free journals & back issues' section on Highwire Press (*http://www.highwire.org*).

Journal quality

When consulting journals you need to have in mind the weight and reputation of the journal. The following 'quality hierarchy' exists:

Review journals
Peer-reviewed journals from
reputable publishing houses
Non-peer-reviewed journals

ISI Journal Citation Reports is a database that can help you evaluate and compare scientific journals. Journals are ranked in subject categories according to how often they are cited in the literature. So, the more a journal article is cited in the bibliographies of other articles, the higher the impact of the journal it is published in. This is called the 'impact factor' of the journal, and its importance is growing in the sciences. You, like other researchers will aim to have your research published in higher impact journals (e.g. the Nature Publishing Group & Cell Press titles).

You can access the ISI Journal Citation Reports through the Web of Knowledge at *http://wok.mimas.ac.uk*. In addition to journal impact factors, the Science Citation Index on the Web of Science (accessed via the Web of Knowledge) gives the number of citations or hits each individual paper has subsequently had; classic or seminal papers will have hundreds of citations.

Copyright

Your institution's library will have paid a copyright fee for its users, allowing copying of articles or books so long as it complies with copyright legislation. For instance, when photocopying from hard copy, a general rule is that for books no more than 10% of a single item may be copied, or one copy of an article from a single journal issue. Multiple copies of articles usually require the authorisation of the copyright holder i.e. the publisher or the author for PhD theses. Some publishers may have special rules; check their websites for details. For electronic journals, it is acceptable to download papers as PDF documents on to your computer to print out, but it is usually prohibited to e-mail these to others. Again, each publisher will have a copyright policy on its website. For a general overview of copyright see the Copyright Licensing Agency website at *http://www.cla.co.uk*.

> **PDF**
>
> Portable document format or PDF is a compressed file format developed by Adobe Systems for displaying and printing documents so that they look the same on the computer screen and in print, regardless of the type of computer and printer used and of the software package that created the original file. The software needed to view PDFs can be downloaded for free at *http://www.adobe.com*.

Unpublished ('grey') material

Documents not widely published – sometimes referred to as 'grey' literature – are often difficult to identify and acquire through mainstream booksellers. The British Library Document Centre has a collection, which is recorded in the British Library Public Catalogue (*http://catalogue.bl.uk*) and is available on inter-library loan. Material coming within this category includes:

- Government publications – most publications are available online from the relevant government departments:
    - Department of Health publications are usually free as full-text PDFs on their website – *http://www.dh.gov.uk*
    - the Stationery Office also lists all government publications – *http://www.hmso.gov.uk*
    - alternatively, the British Official Publications Current Awareness Service (*http://www.bopcas.soton.ac.uk*) provides a subscription-based update of publications and the British Official Publications Collaborative Reader Information Service (*http://www.bopcris.ac.uk*), a free service that searches up to 1995
- Reports – many major organisations in physical and social sciences now issue important information as reports rather than in periodical literature
- Conferences – papers given at conferences fulfil a distinct function in research:
    - you can find conferences and contributed papers on the ISI Proceedings Database (available through Web of Knowledge *http://wok.mimas.ac.uk*)
    - the British Lending Library also holds a comprehensive collection, available to borrow
    - some conference issues of journals contain full-text papers but many only present abstracts. Papers in conference issues are sometimes difficult to locate, so if the ISI Proceedings Database does not index them, you may have to search for a specific website – you can access some conference issues free on the Internet

- Theses:
    - a comprehensive collection of British and Irish theses can be found at *www.theses.com*
    - North American theses feature in the Networked Digital Library of Theses and Dissertations at *http://www.ndltd.org*
    - the British Library also holds some theses as part of its public catalogue
    - a newly developing website for UK doctoral theses is still in its early stages; the project is called Theses Alive – *http://www.thesesalive.ac.uk*
- Patents:
    - the British Library holds a comprehensive international collection of world-wide patents with Internet links to other patent providers – *http://www. bl.uk/collections/patents.html*
    - you can also visit the European Patents website – *http://www.european-patent-office.org*
    - abstracts of US patents are free at *http://www.uspto.gov*
    - Japanese, US, European and World (Patent Co-operation Treaty) published patents (prefixed JP, US, EP and WO) are free, full text, at *http:// gb.espacenet.com* (exact details of the coverage of the espacenet website are available at *http://ep.espacenet.com/espacenet/ep/en/helpV3/coverageww. html*)

Bibliographic databases

Bibliographic databases are likely to be your main reference source as they allow you to search and locate peer-reviewed primary research. These databases provide only the abstracts of articles, so you will need to follow up full-text versions elsewhere. Talking to your institution's librarian will give you all the information you need to locate and access the databases most relevant to your studies. Table 3a contains information on commonly used databases in various scientific disciplines.

The most important database in biomedicine is PubMed (including not just medical journals but biochemistry, cell biology, molecular biology and medical physics). The US National Library of Medicine has made this database free to all at *http://www.pubmed.gov*. Abstracts of papers are usually included, but full-text versions are limited to journal subscribers.

To find papers in the fields of pure chemistry (e.g. *Tetrahedron*) or pure physics (e.g. *Physical Review*) you will need to search the ISI Science Citation Index via the Web of Knowledge at *http://wok.mimas.ac.uk*. The Chemical Database Service, available through the EPSRC Daresbury Laboratory (*http://cds.dl.ac.uk/*), also provides access to various chemistry and medicinal chemistry databases free-of-charge to UK academics; database features include structure-based searching.

Database	Host/provider	Access requirements	Subject covered	Dates covered	Type of material	Search facilities	Download facilities	Other facilities
BIOSIS	EDINA	ATHENS	Life sciences, biology, pharmacology, biophysics	1969 – updated weekly	Journal articles, books, reports, conferences, paper, reviews	Free text, thesaurus	Import filter	Limit, registry numbers
	DIALOG/DATASTAR	Commercial subscription						
CHEMICAL ABSTRACTS CAS–online	DIOLOG/DATASTAR	Commercial subscription	Biochemistry, chemistry, chemical engineering	1907 – updated weekly	Structures, reactions, journal articles, conference papers, patents	Thesaurus, chemical structure–based searching	Import filter	Limit, registry numbers
	STN SciFinder Scholar	Commercial subscription Academic subscription						
THE CHEMICAL DATABASE SERVICE	EPSRC Daresbury Laboratory	http://cds.dl.ac.uk	Various chemistry and medicinal chemistry databases			Chemical structure–based searching		

CROSSFIRE Beilstein	MIMAS	ATHENS	Chemistry, organic	Historical (200 years) – updated weekly	Structures, reactions, journal articles	Free text, chemical structure-based searching		Registry numbers
Gemelin			Chemistry, inorganic					
EMBASE	HILO/KA24	ATHENS/NHS	Pharmacology, biomedicine	1974 – updated weekly	Journal articles, conference papers	Free text, thesaurus, hierarchical	Import filter	EBM filter, limit, registry numbers
INSPEC	EDINA	ATHENS	Physics, electrical engineering, computing, IT	1969 – updated weekly	Journal articles, patents, reports	Thesaurus, hierarchical	Import filter	Limit
MEDLINE (PubMed)	National Library Medicine	http://www.pubmed.gov	Medicine, dentistry, medical psychology, toxicology, genetics	1951 – updated weekly	Journal articles	Free text, thesaurus, hierarchical	Direct to host	Limit, EBM filter, registry numbers
MEDLINE	HILO/KA24	ATHENS/NHS					Import filter	

Continued

FIGURE 3a Table of Major Bibliographic Databases (Continued)

Database	Host/ provider	Access requirements	Subject covered	Dates covered	Type of material	Search facilities	Download facilities	Other facilities
PATENTS	European Patent Office	http://gb.espacenet.com	Life science, physical sciences, chemistry	Historical – present	Patents	Free text		
	US Patent Office	http://www.uspto.gov						
(ISI) WEB of KNOWLEDGE Web of Science: - Science CI - Social Sciences CI - Arts and Humanities CI ISI Proceedings	MIMAS	ATHENS	Biology, chemistry, physics, medicine, biochemistry	1981 (proceedings 1990) – updated weekly	Journal articles, conference papers and proceedings	Free text	Export plug-in	Cited references, limit

Here is a glossary of terms used in the table:

ATHENS – provides users with a single sign-on for nationally provided resources. The most important resources for research students are the online services from Data Service Providers such as BIDS, OVID and MIMAS. Your individual ID/password may not be required for some databases accessed through your institution's network, (will recognise IP address), but will be required for remote access. You can only access those databases to which your institution subscribes

ATHENS/NHS – only employees of the NHS can access this resource through HILO/KA24

CI – Citation Index

Cited references – facility enables you to search for when and where a particular reference has been cited. It is useful to follow up important references or to verify incomplete older references

DIALOG/DATASTAR – commercial database host

Download facilities – this means it is possible to download the reference into your reference manager software, e.g. Endnote or Reference Manager Library (see 'Bibliographic management software' section later in this chapter). You can download references:

- **direct** from the Internet host – this is a straightforward download of the reference from the Internet into your database
- using an **export plug-in** – this means that references from the database are automatically filtered through "export plug-ins", which can be downloaded into your management software directly from the manufacturer's homepage
- using an **import filter** – this means that you need to convert to a text file and then run through an Endnote or Reference Manager import filter

EBM (evidence-based medicine) **filter** – this can be used to select articles where some meta-analysis or research methodology has been applied

EDINA – database host for the academic community

Free text – searching for words in the text of the reference where there is no thesaurus of standardised keywords available

HILO/KA24 (Health Information for London online/Knowledge Access 24) – hosts a set of major bibliographic databases and full text versions of journals and textbooks. Access is limited to employees of the NHS or any student collaborating with NHS employees. At the moment, this service is available only in London and the Southeast of England

Limit – this facility helps you to limit your search for articles to specific dates, data fields or language

MIMAS (Manchester Information and Associated Information) – a national data centre providing the UK further education community with networked access to information resources

Registry numbers – numerical code assigned to chemical substances by the Chemical Abstracts Service

STN – commercial database connection service hosted by FIZ Karlsruhe

Thesaurus – a controlled vocabulary used for indexing articles. It provides a consistent way to retrieve information, particularly where different terminology may be used for the same concepts

How to search bibliographic databases effectively

The availability of bibliographic databases makes literature searching relatively quick. Before you start, take some time to think about your search – develop a plan and a strategy to help you look for information.

Formulate the question in your mind and ensure that you have included all the important factors. Break down your question into the necessary parts so that each term can be entered separately – this will improve the relevance of the retrieval. For example, if you are looking for articles on 'conformal radiotherapy algorithms', this search phrase will only retrieve references where the words are adjacent, and the search result will be very small. How you break down your search depends on the search facilities of the database e.g. whether it is free-text or has a thesaurus.

Searching a 'free-text' database

Introducing operators (Boolean) such as **AND, SAME** or **OR** into your search statement will expand your search and increase the retrieval of relevant references. These are used as follows:

- **AND** operator does not depend on terms being adjacent:

  radiotherapy **AND** conformal **AND** algorithm$

- **SAME** operator will ensure terms all appear in the same sentence and thus increase relevance
- **OR** operator is used to introduce synonyms so that you can retrieve articles expressing the same concept differently:

  radiotherapy **OR** radiation

- **$** truncation symbol to retrieve plurals or derivatives

In the case of the search subject chosen here, the best search statement to use would be:

  conformal **SAME** (radiotherapy **OR** radiation) **SAME** algorithm$

When breaking down your search statement consider the following points:

- first, look for synonyms and search using the **OR** operator
- combine all the concepts using the **AND** or **SAME** operator
- take out any unnecessary terms
- use a truncation symbol to retrieve plurals and other derivations (the symbol may vary between databases)
- check your spelling.

Searching a database with a thesaurus

After breaking down the search statement into its relevant parts or terms, find the appropriate thesaurus term (TH) to match your search terms. Using terms from the database's thesaurus will ensure that you retrieve all relevant articles, no matter how the concept is expressed. In the case of the search subject chosen here, the search statement would now be:

> Algorithm$ **AND** Radiotherapy, conformal (TH)

Some databases have a thesaurus that operates a hierarchy of terms. There are 'preferred' terms and 'sub' terms. When you select a preferred term you will see a list of sub (or related) terms that can be selected and automatically included in the search. Most databases will provide help with searching.

Remember different databases use different terminology. For example, although you can search free-text databases, like Embase, using the term 'cancer' – when you are performing indexed-searching (e.g. in PubMed) you would need to use 'neoplasm'.

Reviewing the results of your search

Once you have run your search, look at the results to see if the search needs revising:

- if you have retrieved more references than you expected, there are several ways to refine your search and make the retrieved references more relevant. Try adding another term or use the 'limit facility' (see database table and bibliography), or create your own limits, by time, publication or title, for example
- if the results are too narrow and too few references have been retrieved you will need to revise your search; this may involve excluding restrictive terms, adding to the fields covered or widening the timescale.

Remember, keep evaluating your references throughout your search – you need to make sure that your results continue to be relevant and appropriate.

Current awareness services

These services are a useful way of keeping up-to-date with the latest literature entered into the bibliographic databases. By creating a search profile, which runs automatically every time the database is uploaded, you can receive details of articles (usually to a registered e-mail account) matching your pre-defined profile criteria. The profile could include criteria such as table of contents, author's name or key words.

Here are the details of the three most comprehensive and popular current awareness services; it is also worth finding out if your institution provides a similar online service:

- ISI Current Contents – this service is available online through the Web of Knowledge at *http://wok.mimas.ac.uk/*
  - literature contained in this service is kept absolutely up-to-date, appearing almost as soon as the journal is published and may even appear before the journal issue is available in the library
  - it is updated daily
  - the search profile automatically selects material according to your submitted criteria
  - abstracts are available
  - it is multi-disciplinary, allowing searching by category
  - it can be accessed through ATHENS
- Zetoc – this service is hosted by MIMAS at *http://zetoc.mimas.ac.uk*
  - very current
  - updated weekly
  - multi-disciplinary, with articles from 20,000 of the most important research journals
  - the search profile is stored with a host so that references are sent by e-mail
  - access available through ATHENS
- PubCrawler – this is an alert service for PubMed and is freely available online at *http://pubcrawler.gen.tcd.ie/*. Pubcrawler is similar to Zetoc and will supply users with updated references and abstracts for tailor-made search terms.

The Internet

The Internet is a vast store of information, with new pages added daily. The sheer size of the Internet can lead you to believe it is the source of all knowledge, especially as it is so quick and easy to retrieve information.

Although it is tempting to rely on direct Internet searching for all your information retrieval, it is worth remembering that:

- Internet search engines look for words or word combinations anywhere on the web page regardless of relevance
- websites can be created without being checked or authenticated, so many can be of dubious value.

There are now numerous Internet search engines you can use to locate websites, the following are the most popular:

- Altavista – *http://www.altavista.com*
- Google – *http://www.google.com*
- Yahoo – *http://www.yahoo.co.uk*

Some broad searches do throw up an enormous number of hits so refined searching is the best way – for instance a search on 'lung cancer' on Altavista produces more than 4.5 million hits.

General Internet searching is best restricted to looking for specific sites or very specific items of information. For general literature searching it is best to consult bibliographic databases or the resources where Internet sites are evaluated and authenticated, for example:

- BUBL (*http://www.bubl.ac.uk*) – this is a comprehensive collection of internet links organised into subject categories which features:
    - selected Internet resources that have all been evaluated and checked
    - information arranged by subject category to include: important sites, mailing lists, e-journals
    - Internet bookshops, e-textbooks
    - information aimed at the subject specialist and the general public
- The Resource Discovery Network (*www.rdn.ac.uk*) – this subject gateway is a collaborative venture with partners such as the British Library and BIOME, along with lecturers and researchers. It pulls together a range of independent information service-providers in many subject areas, including health and the physical sciences. The site also offers free online tutorials in effective Internet searching at *www.vts.rdn.ac.uk*.

The invisible web

Remember that search engines do not pick up on all the available, searchable databases on the Internet. To make your Internet search more effective, add 'database' to your search term. Three such elusive and useful searchable databases are:

- Librarians' Index to the Internet – *http://lii.org*
- AcademicInfo – *http://www.academicinfo.net*
- Infomine – *http://infomine.ucr.edu*

Discussion lists and newsgroups

These online forums or electronic discussion lists enable you to keep in touch with colleagues working in your subject area, find out about meetings, news and jobs. The lists are usually organised on the list-server principle where the host acts as co-ordinator. The lists can cover every possible subject area and have an archive facility enabling you to browse previous e-mails and follow the thread of a particular discussion topic. The main British list-servers are:

- Mailbase – *http://www.mailbase.ac.uk/*
- Jiscmail – *http://www.jiscmail.ac.uk/*
- UseNet – *http://www.usenet.com*

To locate other newsgroups try Google at *www.google.com/grphp*. Other directories of e-conferences and mailing lists can be found on BUBL at: *http://www.bubl.ac.uk/mail/services.html*.

Evaluating information on the Internet

When you are retrieving information from the Internet have a quality check-list in mind:

- what are the qualifications and reputation of the author?
- is the material up-to-date?
- is the material clearly written?
- is the scope of the material clear?
- is it clear who the material is aimed at?
- is it relevant?

Managing your references

You will need to create your own comprehensive research bibliography for your thesis. The best way to do this is to use bibliographic software. Many institutions have Reference Manager or Endnote software for you to collate your references. To familiarise yourself with bibliographic citations, look at previous PhD theses in your academic library to see how references are indexed.

Bibliographic management software

Bibliographic software enables you to create your own database of the references relevant to your PhD. The two most popular examples of this software are Reference Manager and EndNote, both produced by ISI Researchsoft. Both of these have similar functionality and features.

References can be added to your personal database either manually or from online bibliographic databases. Online references can be imported into your database in various ways:

- directly from the Internet host as with PubMed
- formatted automatically through export plug in filters as with ISI World of Science
- downloaded into text (.txt) format and run through appropriate Endnote or Reference Manager import filters as in most databases.

Your own reference library can function like any other bibliographic database and can be searched by author, key word or title, for example. Citations can be retrieved from your personal reference library and then added into your document or thesis. Your document can then be formatted in one of more than 300 styles to create a final manuscript with appropriate in-text citations and reference list at the end.

When deciding which software to choose, consider these important differences:

- EndNote – Windows and Macintosh applications; more information and download trial version at *http://www.endnote.com*
- Reference Manager – Windows only. Multi-user network version available; more information and download trial version at *http://www.refman.com*

For information about organising your reference papers see 'Handling reference papers' in Chapter 1, page 7.

Checklist

✓ Retrieval of information is important to your research directly and also in compiling the bibliography for your final thesis
✓ Databases, software and journals are available to you and major universities will subscribe to most of the main ones
✓ Tutorials and helpdesk information are available on all of the databases cited here, which give details on how to search effectively
✓ The publishing of scientific information is still in flux and may well change in the future if open access becomes accepted

Chapter 4
Critical Reading

by Dr Stan Venitt
Emeritus Reader in Cancer Studies, University of London

Introduction

By far the most important source of information for a research scientist is the peer-reviewed paper containing the results of original research. During your career you might expect to read hundreds if not thousands of such papers. At first the quantity and complexity of the literature may seem daunting, and it can be difficult to know in what depth to read, and which aspects of a paper are the most important. How should you approach this avalanche of reading and make sense of it? The most effective and efficient way is to use the technique of critical reading.

Critical reading means judging the scientific worth of a paper by approaching it in a structured way that allows you to identify the crucial elements of a paper and subject them to your own evaluation.

Scientific papers are written in a formal style and in fairly rigid formats. In this chapter you will find a description of what each section of a paper contains and how to use these sections to focus your reading. Although a standard format is described, some variations may be found between disciplines and also between journals, but in essence the approach from a critical reading perspective is the same.

Scientific papers carry with them the stamp of reliability, honesty and objectivity. These qualities are maintained by the process of peer-review. Before embarking on advice on how to read scientific papers it might be useful to describe how they come to be published.

Peer-review

Publication of scientific papers in peer-reviewed journals is the key process by which advances in science are disseminated to the scientific community. Peer-review simply means that a paper will be published only if it has satisfied at least two reputable anonymous scientists ('referees'), considered by the journal's editor to be expert in the given field. The work described in the manuscript must also conform to standards published by the journal, and to generally-accepted scientific values. Referees usually suggest minor or major modifications that they think will improve the paper. More rarely they will advise the editor to accept a paper unchanged or reject it out of hand. The authors can elect to modify the manuscript in the light of referees' comments and criticisms, and may successfully challenge some of them. However, the editor's decision is usually final, although he or she may seek opinions from additional referees where there is serious disagreement between the original scientists.

Once a paper has been published it is then open to public appraisal and criticism. This is an imperfect system, and unsound claims occasionally appear in print, sometimes with serious consequences. Nevertheless, the system of peer-review has stood the test of time and no-one has yet thought up a better one.

Having gained your PhD, and with some post-doctoral experience, you will be expected to referee manuscripts as part of your duties as a professional scientist. Thus the ability to read papers critically can be thought of as a continual process of peer-review, applied to your own work, to manuscripts received for review, and to papers already published.

Not all sections of a paper will demand your critical attention. Certain speed-reading techniques can be useful for these sections and you can find information about such techniques towards the end of this chapter.

What is critical reading?

Not all the reading you do during your studies and beyond will be critical or in depth. What you read and your purpose for reading will affect the way you read. This chapter is concerned with your primary information source – the scientific paper – and its critical evaluation.

When applied to reading primary-research papers, critical reading means exercising judgment on the scientific worth of the material you are considering. Reading critically is not a negative exercise, designed only to find fault; it simply means that you should not accept what you read at face value. The degree of judgment required depends, to some extent, on why you are reading a particular article. For example:

- reading for interest
- background reading in your subject
- the subject of the paper closely relates to your work
- the paper contradicts your work
- you suspect that the paper contains data or interpretations that you think are misguided or dishonest.

If you are reading for interest, you need not examine every aspect of the paper in order to judge its scientific merit or relevance. In this case, you can skim read or scan the paper for pertinent information. However, if you are reading for any of the latter reasons, then you will need to review critically all the paper's component parts. This requires you to test the author's interpretation of the data and apply your own analysis and evaluation, to arrive at your own conclusion. You may want to reproduce those aspects of the paper with which you disagree in your laboratory.

The act of reading a paper entails judgments that may start to form even before you have read the title. For example, what do you know about the journal in which it has appeared? Is it in the top rank? There is now a distinct discipline called bibliometrics that is devoted to the ranking of journals by a variety of statistical measures, of which 'impact factor' is the one most commonly mentioned (see 'Journal quality' in Chapter 3, page 41; and 'Choosing a journal to publish your paper in' in Chapter 6, page 96). A journal with a high impact factor is thought to be more prestigious than a journal with a low impact factor. Examples of biomedical journals with high impact factors include *Nature*, *Science*, *The Lancet* and *Cell*.

Scientists are often encouraged by their employers or sponsors to submit their papers to journals with high impact factors. This is because organisations that assess the quality of research and award grants base their evaluations not only on the number of papers produced by scientists, but also on the perceived quality of the journals in which they are published. Prestigious journals reject a high proportion of the numerous papers submitted to them and it might be thought, therefore, that the standard of refereeing would be higher for those journals compared with those of more mundane titles. However, a paper might be rejected by a prestigious journal not on grounds of quality, but simply because the editor considers the paper to be too specialised for the journal in question. On the other hand, an editor of a less exalted journal may feel obliged to accept enough manuscripts to ensure that each issue of the journal contains an adequate number of papers. Such pressure might influence an editor to apply less stringent criteria when deciding whether to publish a paper. It is of course impossible to account for all the factors that led a particular paper to appear in a particular journal.

 Remember, however prestigious the journal you are reading, always approach papers with an open, critical mind.

Why is critical reading important?

You need to show clearly that you have read, understood and evaluated the relevant information in your field when you come to prepare your annual progress reports, as well as writing papers and your thesis.

New developments and techniques found in primary-research papers may have a significant impact on your work. For example:

- a new technique may make your work easier, saving you time
- your work may appear to have been duplicated
- information may appear that adds to your work or contradicts it.

On the day of your viva you will be expected to have current knowledge of your subject area. Understanding and evaluating primary research papers is the only way to keep up-to-date with changes in your field.

Many published papers contain errors – these may be factual or statistical, or in the form of the overstatement of the significance of results, or as the presentation of speculation as fact. Deliberate deceptions are rare but not unknown. Only by thorough and critical review of a paper can you hope to identify such errors.

The structure/organisation of a paper

Scientific papers usually conform to a standard format. Here is a description of what each section should contain:

Title	A clear and succinct summary of what the paper is about
Authors and acknowledgments	Who is responsible for the work and their contribution
Abstract	A summary of the principal objectives and scope of the work, the methodology, the results and the main conclusions
Introduction	Provides background information, setting the scene and describes the reasons why the work was carried out. Poses questions/hypotheses to be tested
Materials and methods*	How the work was done. All relevant details of procedure and how the data were analysed. In some journals this appears last
Results	The critical core of the paper. What the researchers actually found. Data can be presented in a variety of ways – tables, charts, photographs – but whatever the presentation, the data should be readily understood and easily cross-referenced to the materials and methods and discussion sections
Discussion	The authors interpret their results in the light of the questions they posed in the introduction and in the context of other data (their own or from other workers in the field). Finally they may draw some conclusions that summarise what they did, why they did it and what it told them
Conclusion	The key discovery or discoveries

Experimental*	Full details of the experimental procedures used to prepare compounds, and the data associated with their identification
Bibliography	List of all the papers referred to in the body of the paper

(*The inclusion of these sections and their position are dependent on the discipline. If in doubt, check with your supervisor.)

Variations on the organisation of a paper

The structure detailed above is a common one but the organisation of a paper may vary between disciplines and journals. In some cases the results and discussion sections are combined, the abstract may provide an introduction or an experimental section replaces the materials and methods section – often in a different location in the paper.

How to read critically

When a primary research paper is of direct interest and relevance to your work you should subject it to critical evaluation. Ask yourself why the work was needed, what the results were, how the author interpreted those results and what your own interpretation is. Before you start, be clear about what you're looking for and where to find it.

Don't read the paper straight through from start to finish – this is a poor use of your time and won't aid your understanding of the work. Certain sections of a paper are more important than others – here is a suggested order for reading the sections of a paper:

1. Abstract
2. Introduction
3. Results
4. Discussion
5. Materials and methods/Experimental

Below is a guide to what to look for from each of the sections. When you approach a paper in this way you are employing a range of skills:

Comprehension	What is the main point of the paper? Have you grasped what the author is saying?
Comparison	How does the paper compare or relate to other work in the same field?

Interpretation	Do the authors' conclusions follow from the results or do you have a different view?
Analysis	Is the paper well structured? Does the evidence support the argument? Does the author make assumptions?
Consolidation	How do the authors come to their conclusions? And how do you draw your conclusions from the results?
Evaluation	Is the work good science?

Of all these skills perhaps the most difficult is evaluation – how to know what is good science. This comes with experience but the questions posed below make a good starting point in the context of critical reading.

Abstract

'Skim reading' (see 'Reading techniques' later in this chapter) the abstract will give you the main points of the work. You need to ascertain what question the paper is asking and what answers it provides. Your first encounter with a paper may be made using an online bibliographic service such as PubMed, where the abstract may be the only way of finding out what the paper is about. However, reading an abstract cannot fulfil the requirements of critically reading the full paper. After you have read the abstract take some time to think about what you know about the subject, then move on to read the introduction.

Introduction

From the introduction you will be able to find out whether the authors present an up-to-date view of the field. Ask yourself whether the authors have described adequately the general and specific background to their study, and the questions their work will attempt to answer. Is the paper now in context?

Results

For the critical reader the results section is the most important part of the paper as it is supposed to present the bare facts. Ask yourself the following questions when considering the results:

• are data from controls easily distinguishable from test data?
• is it easy to identify data that have been mathematically transformed?

- are data from replicate experiments presented separately?
- if the authors employed statistical analysis, has this been presented clearly?
- are there sufficient raw data to allow you to perform your own statistical analyses?

Discussion

Your task is to make your own judgment, based on what you have found in the paper and other knowledge that might shed light on the topic. Question whether the aims of the work have been achieved and the stated hypotheses tested. The authors' conclusions should be clear. Think about whether you agree with the conclusions – do you think they have clearly distinguished between speculation and fact? How does the work in this paper relate to other work in the field? It is sometimes useful to make your own interpretation of the data first and then return to the paper to see what the authors have made of the results.

Materials and methods/Experimental

If you are not familiar with the techniques used, you should scrutinise this section in great detail. You need to be thoroughly familiar with the methodology to evaluate the quality of the evidence. Think about what questions you need to ask yourself when reading this section of the paper – you will get better at this with experience. The questions you ask will depend on your discipline. Those listed below are more appropriate for biologically-focused papers, by way of example. But create your own list and add to it over time:

- is the experimental design sound and is it likely to answer the questions posed in the introduction?
- have the authors described their methods in sufficient detail to allow others to repeat the work?
- have they confirmed their results by performing replicate experiments?
- are there adequate negative and positive controls?
- have they employed appropriate statistical tests?
- what are the limitations to the methodology?

The main sections of the paper for the critical reader have been dealt with above but here are some points to bear in mind about other sections of the paper.

Authors and acknowledgments

It is worthwhile spending a moment or two reading the authors' affiliations, source of funds and statements of possible conflicts of interest.

The order in which the authors are listed indicates the contribution each author made to the paper. The first author listed is usually the one who did most of the work, the rest of the authors being listed in decreasing order of their contributions. However, the final name in the list may well be that of the most senior author, for example, the head of the laboratory in which the work was done. It is customary for one of the authors to be nominated as a 'corresponding author' to whom readers can address queries and comments.

Bibliography

Quickly skim read the bibliography to ensure that the authors have referred to papers appropriate to the immediate field of their study. The absence of important and germane papers in the list of references may indicate carelessness, ignorance or even malice on the part of the authors and should sharpen your critical faculties.

Common problems

- Poor writing – sometimes papers are difficult to understand because of poor structure, an excessive use of jargon or other factors such as the lack of a clear explanation of the significance of the work
- Authors may be uncritical about their experiments
- Authors may not differentiate between facts and speculation – this makes it hard for the untrained eye to know what is actual fact and what is not
- Authors may overstate the importance of their findings

Evaluating your own work

As well as applying critical reading to the papers of others, you need to appraise your own work. When writing your own papers taking an objective and critical view is often difficult.

Tips for critically evaluating your work

- Try to look at your work from different perspectives and identify its limitations, its strengths and its weaknesses
- Is your conclusion clear and unambiguous?

- Ask the same questions of your own paper as you would of papers written by others
- Try to create some detachment by leaving your work for a while – do something else unrelated to your writing and come back to it later
- Ask for feedback from your peers and colleagues

Reading techniques

A useful technique for gaining detailed information and a thorough understanding of a paper is the SQ3R[2] technique. This technique is useful for comprehension and revision rather than critical reading per se, but it can be usefully applied as a way of retaining information when you are reading critically. The technique involves five stages:

Survey – a quick skim read to give you an overview

Question – establish the purpose of your reading. Why am I reading this? What am I looking for?

Read – a slower, more thorough read for comprehension

Recall – take notes or recite the main points and do this regularly throughout the reading process

Revise or review – skim read again and remind yourself of the main points to test your recall

For lighter reading there are some simple techniques you can use to speed up your reading:

- *skimming* – reading to gain an overview of the text, i.e. to find out what the text is all about. Before you start, think of questions such as, what is this paper all about? Is it relevant to me? Ask these questions as you read. Read the text quickly, not word by word. Try reading the first and last lines of a paragraph and look for key words or terminology
- *scanning* – a quick search for key specific points, such as a fact or a key word. To do this you don't need to read word by word, read quickly and stop when you reach the fact or word you are looking for, then concentrate on that section of text so you can understand it in more depth.

---

[2] Robinson, Francis Pleasant (1961, 1970) *Effective Study, 4th edition*, Harper & Row, New York.

Checklist

✓ Before you read the paper ask yourself why you are reading it and what you are looking for

✓ Don't assume that papers published in prestigious journals are above critique

✓ Read selectively and with an organised approach

✓ Do not attempt to understand all the facts. You are looking for the main points in the paper

✓ You may need to re-read the paper or sections of it several times

✓ Don't be put off by what may seem like endless detail

✓ Make notes of the main points as you go through the paper

Finding out more

Books

Bowers, D., House, A. and Owens, D. (2003) *Understanding Clinical Papers*, John Wiley & Sons, Ltd, Chichester.

Cottrell, S. (2003) *The Study Skills Handbook, 2nd edition*, Palgrave Macmillan, Basingstoke. (Contains a chapter on critical reading).

Websites

Collins, L. *How to read a scientific article.* Available at:
http://www.fiu.edu/~collinsl/Article%20reading%20tips.htm

Holmes, M. *Reading skills.* Available at:
http://osiris.sund.ac.uk/~cs0mho/chap11.htm

Hughes, D. (1999) *Effective Reading. University of Bradford Department of Civil Engineering website.* Available at:
http://www.brad.ac.uk/acad/civeng/skills/reading.htm

Huxham, M. and Madden, S. *A guide to the critical reading of scientific research papers. The Talessi project website.* Available at:
http://www.gre.ac.uk/~bj61/talessi/tlr51.html

Larsen, E., Dworkin, I., Cordon, A. and Romans, P. *Primary papers: how to read the paper critically.* Available at:
http://www.cquest.utoronto.ca/botany/bio250y/labs/scien_writ/readpapers. html#critical

Little, J. and Parker, R. (2004) *How to read a scientific paper.* Available at:
http://www.biochem.arizona.edu/classes/bioc568/papers.htm

McNeal, A. *How to read a scientific research paper – a four-step guide for students and for faculty.* Available at:
http://helios.hampshire.edu/~apmNS/design/RESOURCES/HOW_READ.html

SECTION THREE: COMMUNICATION SKILLS

Chapter 5
Oral and Poster Presentations

by Dr Maggie Flower
Senior Lecturer in Physics as Applied to Medicine, The Joint Department of Physics, The Institute of Cancer Research and the Royal Marsden NHS Foundation Trust

Introduction

As a research scientist you will be required to present your work in different settings and in different formats. During your PhD the main way you will communicate your work is through oral presentations, posters, and papers (see Chapter 6 for information on writing papers). You may be presenting to large audiences at major conferences, or to small and informal groups on other occasions. Whatever the setting or format, it is important that you learn how to communicate effectively so that you do your work justice.

When presenting, you have limited time or space to get your ideas across. Your aim is to ensure your work is understood, giving the audience a clear message to take away and think about. So deciding on the content of your presentation, how you want to deliver it and its visual representation all need careful consideration. This task is made easier if you obey certain simple rules and apply the techniques set out below.

Giving a research seminar

When you are a student, presenting a seminar may seem daunting but it is a good way for you to develop your work through discussion with your peer group and colleagues. Giving seminars is intended to help you shape your scientific career, giving you experience of what it's like to work in a research environment. Learning to present your work to an audience also helps you structure your ideas into a form that will be valuable when you come to write about your project.

Preparing your seminar

Planning is the secret of success. This applies to most things and certainly to preparing a research seminar. It is important to start with clear objectives, know what you want to achieve and what you are trying to convey. The following sections discuss the major areas you should consider when preparing your seminar.

Know your audience

It is important to find out what proportion of the audience will be novices or experts in the topic you are going to present as this will affect the content of your seminar.

For example, if the audience is made up of experts, the level of introductory or background information will be lower than if the audience is less familiar with the topic.

The table below gives you an idea of how the content of your seminar should be weighted, based on your audience:

	Introductory or background material	More technical material
Mostly expert audience	50%	50%
Mostly novice audience	70%	30%
Lay audience	90%	10%

Regardless of your audience, your seminar should always begin with an introduction and finish with a summary of key points:

* when preparing an introduction try to make it imaginative and succinct, as this will help to engage your audience
* always allow time for a summary. This is an important part of your presentation, because although the audience's attention may wander during the main body of the seminar, when they hear the words 'Finally, in summary...' or 'The take-home messages are...' their attention will be drawn back to you.

Content

The content of your seminar will be discipline-dependent. Guidelines are given below on the standard outline for the structure of a scientific seminar. You do not need to stick rigidly to these headings but they are listed here in logical order:

Introduction	State *why* you did the work
Materials and methods	Say *how* you did the work and describe experimental procedures e.g. those used to prepare chemical compounds
Results	Outline *what* you discovered
Discussion	Discuss whether your results are what you expected; do they agree or disagree with those of others?
Conclusions	Summarise the key points of your seminar
Future work	What next? Outline your plans for the next experiments you're going to undertake
Acknowledgments	Briefly acknowledge your supervisor and anyone who may have helped you with your work (e.g. scientific officers, postdocs, collaborators)
Questions	It is common to invite questions from the audience at the end of your seminar

Use of maths and equations

Use mathematics and equations sparingly in a seminar. It is not usually necessary to take your audience through the mathematical derivations step-by-step, instead focus on:

- the assumptions used in the derivation
- the techniques used to solve the equation
- the relevance of the equation to your topic (if it's not relevant, leave it out).

Allow your audience time to assimilate the maths; define any unfamiliar symbols and avoid cumbersome notation.

 Remember, when presenting to a lay audience, avoid maths, equations and technical jargon as much as possible.

Length

It is good practice to keep to the time you are allocated for your seminar – at conferences this will be particularly important, so that the schedule for the conference is not disrupted.

Remember, never speak past your allotted time.

Two common questions

How do I present two years of work in ten minutes?

- Any topic can be presented in any amount of time
- It's better to narrow down rather than widen the scope of your talk
- As a general rule use a maximum of *one slide per minute*, even for a short presentation
- This depends on the slide content and length of the talk – 60 slides in 60 minutes, for example, is *not* recommended
- Allow more time for text slides than for images
- If images are used you need to allow enough time to explain the orientation of the image
- When presenting graphs, allow time to explain what the x- and y-axes are, for example, before describing the details of the result

How do I cut down a previous presentation from 30 minutes to 20 minutes?

- REDESIGN your presentation
- Decide on what is going to be removed from your talk
- Condense two slides into one, without compromising their legibility
- DON'T flash through your slides
- DON'T frantically scribble on a whiteboard

Visual aids

There are many factors that contribute to a good presentation and visual aids are definitely one. However, although good slides do not necessarily make a good lecture, bad slides can ruin one.

Microsoft PowerPoint, 35mm slides or an overhead projector (OHP) can be used for presentations. It is up to you to choose which method is best for your purposes. The choice may of course be dictated by the facilities available at a particular conference or venue. You should always check the instructions provided by the conference organisers beforehand. If you have a choice, there are advantages and disadvantages to all three methods, so you will need to consider these when deciding which to use for your presentation.

	Advantages	Disadvantages
Microsoft PowerPoint	• You can incorporate videos and animation into your presentation; these are powerful illustrative tools • It is easy to combine text, graphics and images to make each slide more interesting • You can add elements from other Microsoft applications, e.g. graphs from Excel • Amendments are easily made • Handouts are easy to produce • The bullet-point style makes it easy to organise the presentation of your research material • You can print your presentation onto transparencies as a back up in case of technical difficulties.	• There may not be facilities available to project your presentation from a computer • You may have to take your own laptop • You may have to rely on a technician to run the programme, so you will have less control • Technical problems may occur • You may need to allow more time between presentations for setting up (e.g. if several people are using different computers)
35 mm slides	• It is possible to produce slides of excellent quality • Slides are more professional than hand written transparencies • Slides are universally accepted at scientific meetings – although this is changing with the wider availability of computer facilities • Slides are easy to store	• The speaker has less control – you may have to rely on a projectionist to change slides • Projectors can be noisy/jam/obstruct view • Eye contact with the audience can be more difficult • You need to prepare material well ahead of time • Any amendments will involve making new slides • Slides are expensive • Technical problems may occur
Overhead projector	• OHP transparencies are cheaper than 35 mm slides • Transparencies are more informal and useful for small groups • It is easy to face the audience and maintain eye contact • You can build up information with overlays • You can use coloured pens to highlight different things • You can add information by hand during your presentation – but don't scribble!	• You may obstruct the view of the screen while you're speaking • Depending on the quality of the projection equipment, the projected image may be distorted/blurred • Transparencies often attract static and don't stay flat

Dual projection

There may be facilities to project two 35 mm slides at the same time, on separate screens. This can be useful, but don't be tempted to cram in twice as much information. Try a combination of text and graphics e.g. text on one slide, graphics on another, but make sure that text and graphics correspond. You could also use an OHP for text and 35 mm slides for images or photos.

Practising your presentation

Giving yourself the opportunity to run through your presentation beforehand is an essential part of preparing your seminar. It can highlight omissions or mistakes, and also increase your confidence when you come to present on the day.

Rehearsing in front of colleagues or friends is particularly useful for your first few presentations, until you start to gain more confidence. You can either run through your talk as you would on the day, or ask them to interrupt with tips as you go along, depending on how much time you have before your presentation day.

Thinking through your talk is NOT sufficient. You need to practise so you can both duplicate the stress felt in front of an audience and judge how long your presentation will be. As you stand in front of an audience, you will find you talk more quickly, so the presentation might be shorter than you expect. Speak loudly and clearly – this will prevent you from speaking too fast if you are nervous. Learn the order of your slides as this will prevent you from looking surprised as the next slide is displayed. Also practise answering questions from your colleagues; think in advance about what questions you may be asked.

A rehearsal on your own in the room where you are going to give your presentation is useful too. It helps to familiarise yourself with the acoustics and technical aspects ahead of time. During rehearsal, check your slides for:

- timing (how long it takes you to present each slide)
- legibility
- correct order and orientation
- moisture on 35 mm slides (any condensation will be magnified when projected)
- thumb prints on transparencies or 35 mm slides.

Tips for the preparation of your presentation
Regardless of the methods you use to produce and project your slides (or transparencies), the key thing is to make them as legible as possible to

everyone in the auditorium. As you produce your presentation, try to stick to the following guidelines:

- Text:
    - don't use full sentences . . . but don't 'over-abbreviate' either
    - DON'T USE CAPITALS FOR THE WHOLE SLIDE . . . but DO use CAPITALS for emphasis
    - use a maximum of six words in the title, seven lines of text per slide and seven words across the slide
    - don't use too many colours or a complicated background
- Tables:
    - tables should have no more than four columns, and no more than seven rows
    - in general, most tables produced for journals should be redesigned and simplified for presentation at a seminar
- Graphs and charts:
    - use horizontal labels on graphs, even for the y-axis
    - minimise the number of tick marks and number labels
    - all legends, captions and axis labels should have an adequate font size. Use the '1/50th' rule for the minimum font size, i.e. if the long axis of the graph is 20 cm when printed, the smallest letter should be 4 mm (or 20-point font size)
    - if there are multiple lines on a graph, ensure they are clearly labelled and easily distinguishable
    - don't use 3D if 2D will do – sometimes using an elaborate 3D representation can make the graph or chart harder to understand
    - histograms or pie charts are better than tables
    - use colour to highlight parts of a complex diagram
- Animation and illustrations in PowerPoint:
    - animation can be useful for emphasis
    - but don't use it too much as this can be distracting and confusing for your audience
- Duplicate material:
    - if you need to refer to the same slide or transparency more than once, in different contexts throughout your presentation, it's better to duplicate the material. This will prevent you from having to search through the slides you've already shown during your presentation

Presenting your seminar

There is a lot to think about when presenting your seminar, particularly remembering what you're going to say at a time when you will be feeling nervous. Here are a few things to consider to ensure that the presentation runs as smoothly as possible.

Controlling nerves

Even the most experienced presenters get nervous before giving seminars or talking to conference audiences, so you're not alone. Being nervous is a natural response and one that actually works to improve performance. It is simply a part of delivering a good presentation and can help to keep you on your toes. Try not to worry too much about nerves you may feel before you start. The benefits of being well prepared and well practised cannot be underestimated and will help reduce nerves on the day.

There are some strategies to help you if you do get very nervous. Use positive self-talk – this is useful in many situations not just when presenting. Tell yourself that you can do it and remind yourself that nothing 'bad' is going to happen as a result (see 'Using self-talk to develop assertiveness' in Chapter 2, page 22). Here are some key points:

- take a deep breath before you start and concentrate on the key message that you want your audience to take away
- some people find that memorising their opening sentence is a useful way to get the talk started
- think about your body language; if you have a tendency to gesticulate, think about what you're going to do with your hands
- try not to panic, speak slowly, keeping your head up
- and finally, dress appropriately so you feel confident and comfortable when you stand up to speak.

Before and during your presentation

Prior to giving your presentation it is useful to:

- find and familiarise yourself with all the control buttons
  - light switches
  - slide controls
  - microphone
  - one or two buttons for dual projection of 35 mm slides
- find and familiarise yourself with the pointer
- think about where to stand, especially if you are using an OHP or computer
- test the lighting levels for all colour slides, especially images and photos.

Tips for using PowerPoint

- Control the slideshow yourself:
    - you can set the programme to scroll through the slideshow itself using a timer but it is best to switch this option off. If you take a bit longer to do one slide than the next, or if you're interrupted, you will lose the thread of the talk
- Use a pointer where appropriate:
    - you are usually given a laser pointer at conferences
    - DON'T wave the pointer around the lecture room
    - the computer mouse can also be used and may make it easier for you to maintain eye contact with your audience
- Try to prevent a 'shaky' pointer:
    - rest your arm on the podium
    - support the pointer hand with your other hand
    - switch the pointer off when not in use

Tips for using 35 mm slides

- Make sure your slides are in the correct order and orientation:
    - this will need to be done when you load the projector, before you start your talk
    - a red dot is usually placed on the bottom left hand corner of the slide mount. When the slides are placed in the carousel the slide is rotated through 180 degrees so that all the red dots can be seen and this confirms that the orientation of the projected slide is correct
    - make use of facilities at conferences, where a slide preparation room is often provided for your final checks and final practice
- Use a pointer where appropriate ... see PowerPoint tips

Tips for using an OHP

- Stand at the SIDE of the screen:
    - stand next to the projector to avoid getting in the light
    - you may sit next to the projector but this is more informal and should be used only for tutorials or small informal groups (see 'Teaching small groups' in Chapter 2, page 30)
- Check each transparency is projected onto the screen the correct way round before you start talking about it
- Use a pencil as a pointer:
    - point to information on the projector itself; this is easier than pointing at the screen and enables you to remain facing towards your audience
- Cover up part of the text if necessary:
    - using a piece of paper to do this can work well

- Ensure you have room on a table adjacent to the projector for your notes and transparencies
- Ensure the transparencies are in the correct order:
    - you can keep them in a ring-binder and take them out as and when you need them
    - number the transparencies inconspicuously

Delivering your presentation

Here are some rules you should aim to follow when delivering your presentation. You could add to this list when you attend particularly good (or bad) presentations.

Do ✓	Don't ✗
✓ Take a few deep breaths before walking to the podium to calm your nerves	✗ Panic ✗ Mumble
✓ Face the audience	✗ Avoid eye contact
✓ Smile at the audience and the person introducing you when you start your presentation	✗ Scowl
✓ Speak slowly and clearly	✗ Appear nervous
✓ Project your voice (if no microphone)	
✓ Be calm and in control (stand tall!)	✗ Read from written text
✓ Use slides as your notes (don't read from a script)	✗ Pace up and down ✗ Fidget with the microphone
✓ Talk through equations and give the orientation of graphs and/or images	✗ Jangle keys or change in your pocket ✗ Tell jokes that could fall flat
✓ Establish eye contact with a few friendly faces	✗ Ignore your audience (or some sections of it)
✓ Make use of both positive and negative feedback from the audience	✗ Stare into space above the audience
✓ Keep the audience awake by:	✗ Turn your back on the audience
• interleaving images or diagrams with text	✗ Focus on:

• handing round hard copy images or brochures	• your slides or transparencies
	  • the floor/your shoes
	  • one particular person

• showing small pieces of equipment	✗ Give out copies of your slides before your talk: • this will distract the audi-

Do ✓	Don't ✗
• inviting audience participation (but only if you allow time for this) ✓ Carry on talking if the computer or slide projector fails: • for this purpose take to the podium a list or copy of your slides and/or a set of cards with your key points	ence from what you're saying ✗ Go out drinking heavily the night before a presentation: • this will increase your chances of getting 'the shakes'

Answering questions at the end of your presentation

It's common to allow time at the end of your talk to answer questions from the audience. The organisers of the conference or seminar will usually specify the amount of time available.

By way of preparation it is always useful to anticipate what questions might arise and what your answers might be.

Tips for answering questions

- Listen carefully!
- Don't interrupt the questioner in mid-sentence
- Be prepared to rephrase their question:
    - this gives you time to think of your answer and a chance for people at the back to hear the question
- Keep your answers short
- Confess your ignorance, if you don't know the answer:
    - don't try to bluff your way through an answer, remember you may be talking to people who have more knowledge of the subject than you
    - refer the question to your supervisor or another colleague if they are present, but make sure you agree this with them first
- Deflect hostile questions:
    - these occur rarely, but if a questioner appears to be hostile, do not rise to the bait and have an argument in public
    - suggest politely that you continue the discussion after the seminar

Remember, keep a note of the questions asked of you at seminars, similar ones may come up in your viva.

Preparing a poster presentation

You may be asked to display your research findings in poster form at a conference or event. A good poster can be just as effective as an oral presentation, so it's worth preparing well. When you are preparing, remember that your poster is presenting the highlights of your work.

Preparation

Before you start, always read through the information provided by conference organisers. Here you will find any specific requirements of the meeting and the size and orientation of your poster board. It may be useful to mark out this area when you are planning, to get an idea of space available.

You will need to think about how you want to present your poster. For example, it could be a series of A4 sheets (often mounted on card and laminated) or a printed, glossy poster. Talk to people in your lab about the facilities available to you and about time and costs.

You then need to work out the scientific content. Re-read the abstract you submitted and ask yourself:

• are those statements still accurate?
• what data do you need or want to include to illustrate your findings?
• what are the key points you want to communicate?

 Remember, your poster should be a stand-alone, self-explanatory representation of your work that is relatively simple and easy to follow.

Structure and design

Posters are a visual communication of your research so try to keep text to a minimum. Use graphics, such as photos, figures and tables to 'tell your story'. Avoid over-complicated images; your findings need to be clear and also visible from a short distance away.

Try to guide your audience through the research by presenting information in a logical sequence, using arrows or numbers to direct them, for example. Typical content and layout for a poster are shown in Figure 5, you do not have to follow this exactly.

FIGURE 5 Typical poster layout

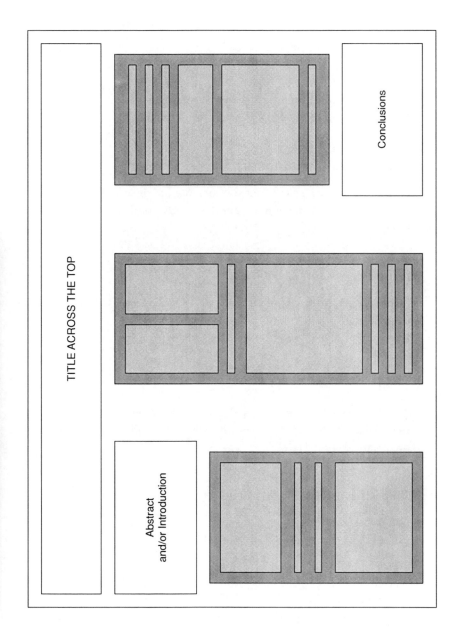

Title

- The title should be short and attention-grabbing if possible
- It should be clear from a distance of three metres
- Use bold, black typeface (about 24 font size; author names should be slightly smaller)
- Include your university, institute or sponsor's logo in this space if you wish

Poster number

- Check where you are to display your poster number
- It is also common to include it at the start of the title

Abstract and introduction

- Display the abstract exactly as it was submitted to the organisers
- Include a brief introduction to your poster or work if you think it adds something

Methods

- Depending on your discipline, this can be headed materials and methods or experimental
- Keep this brief, e.g. just a few lines for each experiment

Results

- The results should form the major part of your poster
- Ensure that graphs and charts are self-explanatory and keep additional text to a minimum

Discussion and/or conclusion

- Be brief
- Present these as a numbered or bulleted list

The visual impact of your poster is important. Avoid clutter – a clean, simple design is most appealing. Also, think carefully about your use of colour and how you are going to arrange the information before you commit to a design. Here are a few ideas:

- background – try a single colour or two to three related colours for different sections (muted shades are often best)

- areas of white or empty space can be used to differentiate elements of the poster
- vary the size and spacing of sections to add interest (but don't overdo it)
- outline or alter the background to graphics for emphasis – dark images look good against pale colours and vice versa; a neutral background can emphasise colours in an image whereas a pale one will reduce their impact

Remember, at least one author should be available during the display session to talk about the work in more detail.

Tips for poster presentations

- Present the highlights of your work
- Make information clear and only as complex as it needs to be
- Keep text to a minimum
- Make the design interesting and not too cluttered or over the top – remember that in the end your science *is* the most important thing
- Give credit where it is due (contributors, sponsors)
- Create an A4 handout – mini-version of the poster, collection of Power-Point slides or brief summary

And finally...
When you attend conferences and meetings remember to look out for good and bad presentations and posters. It is best to learn from the mistakes of others rather than from your own.

Checklist

✓ Plan the content of the presentation well in advance
✓ Obey all the rules regarding content and layout of slides and posters
✓ Rehearse to check timing and content of your seminar are as planned
✓ Obey all the rules to be remembered during your presentation
✓ Remember to take your poster/presentation with you to the conference (preferably as hand luggage)
✓ Re-check your session time and date

Finding out more

Books

Cryer, P. (2000) *The Research Student's Guide to Success, 2nd edition*, Open University Press, Buckingham.

Websites

Chapman, D. (1988) *How To Do Research at the Massachusetts Institute of Technology, Artificial Intelligence Lab: Talks.* Available at:
http://www.cs.indiana.edu/mit.research.how.to/section3.8.html

Moss, H. (Department of Experimental Psychology, University of Cambridge) *Show and Tell: Presenting your research.* Available at:
http://www.bio.cam.ac.uk/gradschool/current/courses/moss/presentation_slide1. html

Radel, J. (1999) *University of Kansas Medical Center online tutorial series: Effective presentations.* Available at: *http://www.kumc.edu/SAH/OTEd/jradel/effective.html*

Chapter 6
Writing a Paper

by Dr Jeff Bamber

The Joint Department of Physics, The Institute of Cancer Research and the Royal Marsden NHS Foundation Trust

Introduction

Scientific papers allow scientists to read about the latest research and discoveries within a particular area. Getting papers published is an integral part of the research process, since the point of research is to contribute knowledge to the world. Therefore, without any form of publication, research projects remain incomplete and, in effect, worthless. During your time as a PhD student you may publish your own work in a journal. Even if you don't publish at this stage, you will need to if you continue working within research. Without published papers, you cannot gain a reputation as a scientist and you will be unable to win grants to fund your work.

The aim of this chapter is to give you some guidance on how to write a scientific paper for publication and to offer some advice on submitting your paper to a journal.

You may be aware of a debate about the advantages and disadvantages of publishing before submission of your thesis. But, working on the assumption that you do not devote excessive amounts of time to papers for publication, to the detriment of your studies, then the overall view is that getting papers published is advantageous.

And that generally the more practice you get the better.

What is a scientific paper?

Scientific papers are a means of disseminating information and new discoveries within the scientific community. They are written, published reports describing original research results. The format and process of paper publishing have been defined by three centuries of developing tradition, editorial practice, scientific ethics and the evolution of printing and publishing procedures. Their existence and accessibility is very important for the development of science and medicine.

Published in a permanent form

No matter how good your work is, it will not have an adequate impact if it is not published in the right form. This means that it must be published in the right place – usually a primary journal, preferably one that is widely read and highly respected in your particular field. Your supervisor and other senior members of your department will be able to advise on this. There is also an online list of journal 'impact factors', which is used by funding authorities, potential employers and other scientists when judging your work and track record (see 'Journal quality' in Chapter 3, page 41).

It is important that others can cite your work. For this to occur, it has to be published in a permanent form, be available to the scientific community and be accessible to information-retrieval services (see Chapter 3). It should contain sufficient detail to allow the reader to repeat your experiments and test your conclusions. Finally, it should be peer-reviewed, a process of quality control operated by respected journals but not, for example, by most conference-proceedings editors (see 'Peer-review' in Chapter 4, page 55).

Impact of the web

Many journals are now available online, as well as on paper, a trend that is likely to become increasingly common. This can affect the way a paper is submitted to a journal and the speed of publication. It also means that readers can view the work the minute it has been published, and not have to wait for the issue to arrive by post or appear in the library.

It is increasingly common for research work to be found on private and institutional websites. While this is a useful additional form of dissemination, it is not a substitute for journal publication. Work appearing on such websites falls short of the requirements for published work because there has been no peer-review process, so there is no guaranteed permanence, and the work will not be accessible via academic information-retrieval services. For the same reasons, it is not advisable to cite references in this form, unless you have no alternative.

Definition of a scientific paper

The Council of Biology Editors' definition of a 'scientific paper' sets out the requirements for published work:

'An acceptable primary scientific publication must be the first disclosure containing sufficient information to enable peers:

1. *to assess observations,*

> 2. *to repeat experiments, and*
> 3. *to evaluate intellectual processes.*
>
> *Moreover, it must be susceptible to sensory perception, essentially permanent, available to the scientific community without restriction, and available for regular screening by one or more of the major recognised secondary (bibliographical) services.'*

Secondary publications, for example, review articles, play an important role in providing an overview of a subject area.

Getting started

Ideally, you should start writing your paper before you begin any experimental work. You may find it helpful to sketch out a plan – this will make it easier to follow a logical order and to write your paper in full. It can also be useful to write the introduction when you are planning your experiments. This ensures that the context of your work remains in focus. Remember to also consider which statistical analyses you will apply to your data as this will often inform and influence the planning stage.

Organisation

Being organised is the key to writing a good scientific paper. You will need to deal with a large amount of data and a substantial quantity of literature. So, your work must be divided into defined sections to ensure it is clear, concise and understandable. It should be easy for someone else to read your research paper and to understand its significance.

Know your audience

The audience at which you are aiming your paper is very important. Do you want your paper to be read by a particular specialist audience, for example, or a more general audience? It is vital to get the emphasis right when you begin writing. For this same reason, it is a good idea to have in mind one or two journals in which you would like to publish your work.

Background reading

You should be reading scientific papers before you start to write your own. This will help you to establish the significance of your work and also give you some guidance on how to present your findings. It will also help familiarise you with the editorial requirements of different journals. Remember that all published papers have undergone a rigorous peer-review process during which the authors have probably had to make revisions.

Standard format of a paper

Scientific papers usually follow a standard format as detailed in Figure 6, below. You should remember, however, that this structure may vary between journals and disciplines.

Although this format has been followed for a long time, it wasn't until 1972 that The American National Standards Institute officially declared it standard. It is important to follow this format. Not doing so makes it harder for other scientists to read, understand and assess your work, and therefore potentially reduces its impact.

How long should a paper be?

The rule is to be as concise as clarity permits. The paper should be as short as possible but each of the sections of the paper must fulfil certain requirements, as described below. Look at some example issues of your intended journal to get an idea of the typical length of papers that it publishes and, if possible, aim to

FIGURE 6 Standard paper format	
Section	*Question answered by section*
Title	
Abstract	
Introduction	What was the problem and why was it worth studying?
Materials and methods*	How did you study the problem?
Results	What did you find?
Discussion	What do these findings mean?
Conclusion	What was the key discovery?
Experimental*	What was the characterisation of the chemical compounds prepared?
References	

(*The inclusion of these sections and their position are discipline dependent; if in doubt, check with your supervisor.)

match the shorter end of the spectrum. Other scientists are more likely to give a short paper a thorough reading, which will give it greater impact. Also, you will want your paper to have as smooth a ride as possible when passing through the peer-review process, so that it is published quickly. The psychological impact on a reviewer when a short paper arrives is not to be underestimated. Short in this respect varies with the subject area, but is probably in the region of ten pages of double-spaced text combined with five figures – twice this length of text and figures would create the impression of a long paper.

Giving your paper a title

Accuracy and brevity are the watchwords here. The title is the most widely read part of your paper and gives the reader the first clue as to its content. It is therefore essential that the meaning is clear. Similarly, in terms of literature searching, abstracting services depend upon the title's clarity and accuracy; any ambiguity could lead to your paper getting 'lost in the literature'. For you, the title gives direction to your writing. Referring back to it ensures that you maintain the emphasis of your paper and do not get side-tracked.

Ensure your title is:

- accurate – choose the words carefully and make sure they are appropriate
- succinct – describe the contents of your paper in the fewest possible words
- clear – good syntax will convey the correct meaning
- a summary of the main message of your paper
- free from abbreviations, jargon and non-essential words.

Listing the authors and their addresses

The order in which you list the authors is important for the assignment of credit. Generally, the first author listed should be the author who did most of the research and writing for the paper. The name of the leading associate should appear second and the third author should have a lesser role, and so on. The 'leading associate' normally makes a substantial contribution to the research and/or writing. Exact author contribution can sometimes be difficult to assess and needs to be discussed between co-authors.

Be aware that listing conventions may vary between journals and research labs, especially between those in different countries. For example, heads of some research labs or groups often prefer to be the final author. Be sure to check first.

The authors' addresses should be clearly identifiable and appear in the same order as the authors themselves. A code (similar to a footnote) can be used when there are, for example, four authors from two institutions.

Citing the literature

There are some golden rules to follow when citing the literature:

- whenever you draw upon information contained in another paper, you must acknowledge the source:
  - all references cited in the paper must be listed at the end of the paper
  - do not list papers that you have not cited but just read along the way – there should be no redundant references
- avoid, if possible, citing references that will be difficult to obtain, such as unpublished data and theses
- avoid using secondary references:
  - if you do use references that someone else has used, always check the original source to ensure they are correct
- the reference should be placed at the point in the sentence to which it applies and not simply at the end of the sentence
- be consistent with style:
  - journals vary considerably in their style of handling references, so consult the journal you wish to publish in first to see how they do it
  - usually the list of references will be ordered either alphabetically by author's surname or by number
  - each reference usually contains details of the author(s), title of the article, name of the journal or book (and place of publication for books), year of publication, volume and page numbers
  - be consistent with the use of journal abbreviations
- references should be perfect so always check all parts of the reference against the original publication – editors do not take mistakes lightly

Preparing the abstract

The abstract is a summary of the work reported in the paper. It allows readers to identify articles of interest to them quickly and accurately, so they do not have to plough through the whole document first. Together with the title, the abstract is also used by databases and is important in literature searching.

The abstract should almost invariably be written last. It should briefly state the principal objectives and scope of the investigation, describe the methodology employed, summarise the results and state the main conclusions. Most journals specify a word limit, typically not exceeding 250 words.

 Remember, do not go over the word limit!

Tips for writing an abstract

- The abstract should never contain information or conclusions not given in the paper
- It should be understandable by itself as it will be published separately in abstracting services
- References should not be cited:
    - leave out references to literature and to figures or tables
- Leave out any obscure abbreviations
- You can add a keyword list at the end of your abstract, which is used by indexing and abstracting services:
    - by adding keywords not already in the title or abstract, readers may find it easier to locate your paper
    - the list should contain the most important words relating to your paper

Writing the introduction

The principal function of your introduction is to establish the significance of your work. Use it to give a background to your research, review the relevant literature and provide a rationale for your study. The reader will want to know why there was a need to conduct the work.

Aims and objectives

The introduction must clearly state the aims and objectives of the work. It may help the clarity of your description to state the long-term aims first and then the specific objectives of the study reported in the paper. This way you increase the reader's interest and set the work in its proper context. There are various ways to state the objectives, such as:

- a research hypothesis to be tested
- a research question to be answered
- another specific objective to be reached, such as the development of a satisfactory experimental or computational procedure with a stated required performance.

Consider which is the most appropriate for your purposes. In the conclusion to the paper you will give the principal finding, for example, whether the results were consistent with the hypothesis or what the answer to the research question was. In the results section, you will be providing the evidence that allows you to draw this conclusion. These factors may influence your choice of method for stating the objective. Try to keep to one principal objective for the paper. If you have more than one objective, make sure that the relative importance of each is clearly stated in the introduction.

Method of investigation and the main findings

Next, you should briefly state the method of the investigation, and sometimes the reason for choosing this method.

It is common to end the introduction with a short statement of the main findings of the paper. You don't need to keep your readers in suspense. Many authors do not disclose their important findings until the end of the paper, but it can be more useful to let your readers know these from the beginning, ensuring that they remain interested in your work.

Writing the materials and methods/ experimental section

The main purpose of this section is to provide enough detail to permit another scientist to duplicate your work and check your study. For your work to have scientific merit, others need to be able to reproduce your results. You should give clear descriptions of materials and equipment available off the shelf, including the name of the supplier where appropriate. You often don't have to detail the preparation of standard mixtures and compounds or routine methods. Instead, if some of your methods are already published, you can use references.

It is usual to present your methods chronologically. However, related methods may need to be described together, so strict chronological order cannot always be followed.

You may want to use illustrations to improve the clarity of your methods. Make sure they enhance the reader's understanding and don't detract from it. Methods need to be easy to interpret, and clearly labelled and captioned.

Experimental section

In chemistry and medicinal chemistry papers the 'experimental' section is usually placed at the end of the paper and replaces the materials and methods section found in the papers published in the biological and physical sciences. It details the methods of preparation of the chemical compounds described. It will include the data (physical, spectroscopic, analytical) that allows the structure of the compounds to be identified, and assesses the purity of the compounds. Most journals publishing papers with experimental details will require authors to meet minimum standards for the nature of the identity/purity data, specified in the instructions to authors.

Presenting the results

The results section should comprise two parts. First, it is useful to restate your experiments, but without the detail presented in the materials and methods. This will help to reinforce the reader's understanding of how the results are organised. Second, you need to present your data.

Data should not be published raw, but should be digested and condensed – you don't need to publish every piece of data, but if you refer to an experiment without including the data, note in the text 'data not shown'. Try to discriminate between what is important and what is not, and present representative information rather than repetitive data. Important trends should be extracted and described in this section.

Use of tables and figures

Tables and figures are central to the results section, especially for large sets of data. Think carefully about what to present in tabular form and what to present graphically. You should choose images and graphs to illustrate a defined point. If you want to find out more about the use and preparation of tables and figures, there are some tips in the following dos and don'ts table.

 Remember, data that can be described in a sentence or two can be put in writing. However, data should not normally be repeated more than once (e.g. not in both the text and in a table). The text in the results section should help to orientate your reader, leading them from item to item, and drawing attention to the key findings shown in the figures.

Dos and don'ts for preparing tables and figures

Do ✓	Don't ✗
✓ Include tables and figures that are necessary	✗ Add illustrations not referred to in the text
✓ Present data in text, table or figure	✗ Present the same data in more than one form
✓ Add concise and self-explanatory captions and legends	✗ Use superfluous words in headings

Do ✓	Don't ✗
✓ Number tables and figures	✗ Include columns of data with the same number throughout: • if the value is important, use a footnote instead
✓ Choose units of measurements to avoid an excessive number of digits	✗ Extend the axes of graphs far beyond the data range
✓ Leave sufficient room for labelling	✗ Fill an entire page with the table or figure if this will leave little room for headings and axis numeration
✓ Make graphs and images large enough to read and interpret	✗ Use small or fancy fonts

Tips for using tables and figures

- Include only tables and figures that are necessary:
  - don't include any that aren't referred to in the text
- Make sure that the figures and tables are clear, legible and simple:
  - each should be self-explanatory from its caption or legend
  - captions should be at the head of the table, legends below the figure
  - headings of table columns and axes of graphs should be brief but informative (include all units of measurements)
  - don't try to cram too much information into one figure
- Consult the guidelines given by the journal you wish to publish in:
  - conventions for producing tables and figures may vary between different journals
  - some journals may not want you to use colour
  - some journals offer guidance on the desired final size of reproduction of each figure

Use of statistics

Here are some guidelines to follow:

- don't just give a numerical value without an explanation of what it is
- for example, you should state whether it is the mean, median, range or another parameter
- include a reference to the literature for a previously published statistical analysis
- for any new methods, provide the detail to repeat them.

Also see 'Statistics packages and image file formats' in Chapter 7, page 109.

Writing the discussion

The discussion section examines the interrelationships between the facts uncovered and the significance of the findings. You should briefly restate your main results, but you also need to interpret them for the reader. It is useful to discuss what principles have been established or reinforced, but you shouldn't extend your conclusions beyond those directly supported by your results. The discussion is also the place to mention any implications of your work.

Consider the following points when you are writing the discussion:

- present the principles, relationships and generalisations shown by the results
- point out any exceptions or lack of correlation
- compare the results with the theory – this may produce further results as a consequence. Such results might, for example, be measures of agreement between experimental results and theoretical predictions, leading to new findings. Whether you report such comparisons in the discussion or in a second part of the results section will often depend on what you want the main focus of the paper to be
- state any relevant points that your study didn't address or which remain unsettled
- show how your results or interpretations agree or contrast with previously published work
- discuss practical or alternative applications of your work
- discuss theoretical or practical implications of your research
- consider the relative merits of other ways the experiments could have been done
- explain why the results can be believed and/or where confirmation is required.

Writing the conclusion

The conclusion is the place to summarise and emphasise your principal findings. It is useful to rank your conclusions in order of importance – the first and last paragraphs should be kept for the most significant findings, with the less important ones in the middle. Your discoveries should be linked to the original problems identified in your introduction, in terms of the hypothesis tested, the research question(s) that the study was designed to answer, or whether the objective was reached.

You should distinguish this section from the discussion, which is the where you speculate about the broader implications of the results.

 Remember, do not draw conclusions that are not supported by evidence presented in the paper.

What to do if the writing stalls

Even the most experienced authors sometimes get writer's block. If you get stuck when writing your paper, there are some things you can do to get back on track:

- get feedback from colleagues:
  - talk to others about your paper as you write it, so you can get feedback and ideas. Also, show the written work to your colleagues, but make sure you have proof-read it first
- reorganise the paper:
  - consider whether there are better ways to order the methods and results; using the order in which experiments were done doesn't always provide the clearest description, so think laterally. Are you trying to say too much in one paper? Perhaps you have enough material for two?
- have a short break:
  - try putting your paper aside for a day or two. When you re-read it, consider whether the original concept for the paper (embodied in the title and described within the introduction) accurately describes what you have done. If it doesn't, start again from the beginning, using the first draft as material you can cut and paste into the new version

Style and formatting conventions

Your main priority is to write your paper clearly and concisely. To achieve this it may help you to aim at an explanatory or tutorial style. You should try not to use more words than necessary and should avoid using jargon. It is also good practice to devote each paragraph to a single point and make one paragraph flow logically to the next using linking sentences.

By all means try to make your paper interesting, but remember that interest in scientific reporting comes from the science more than the style. Interesting style can conflict with clarity if you are not very careful. This is because in literature the thing that captures a reader's interest most of all is ambiguity, but ambiguity represents the antithesis of good scientific style.

Many journals now provide electronic templates for the preparation and submission of manuscripts, ensuring that the layout and formatting corresponds to the 'house style' of the journal.

Using language

There is a trend towards the active voice in modern papers, and you will find that this can be more precise and less wordy. However, some journals and disciplines still prefer the use of the passive voice. It is advisable that you check with your supervisor or your journal of choice.

Jargon

Jargon falls into two categories:

- confused or unintelligible words, or obscure language
- technological terminology or language specific to an activity or profession

Confused or obscure language should always be avoided but in scientific writing some technical terminology is inevitable. Remember, if you use unusual technical terminology, you should define or explain the word first.

Which tense to use

For most of your writing, the past tense is appropriate. However, when you introduce tables, figures or results of calculations, the present tense should be used. For example, it is correct to say, 'Table 1 *shows* that. . . ' or 'these values *are* significant. . . ' try to not switch between tenses unless there is a clear reason for doing so.

 Remember, be consistent with tense.

Using abbreviations and units

Try to keep abbreviations to a minimum, as they can make it difficult for an editor or reader to follow your work. Do not use abbreviations in the title of the paper, and almost never in the abstract – abstracting and indexing services discourage it. Only if you use the same long name or word a number of times should you include abbreviations in the abstract, and then it must be defined on first use.

It is acceptable to use abbreviations elsewhere in the paper, but consider the following:

- if you do shorten a word, you should spell out the word in full first, with the abbreviation following in brackets
- don't use an abbreviation when a word is only mentioned a few times in the paper
- don't make up your own abbreviations (use standard ones), unless a word is used a lot or is very long.

Units of measurement need to be abbreviated when used with numerical values, but are not abbreviated when used without numerals. Don't begin sentences with numerals or abbreviations. It is common practice to spell out in words numbers less than 10, and use numerals for those numbers greater than 10.

Remember, be consistent with abbreviations and units.

Tips on formatting
- Always pay careful attention to the layout of your paper:
    - ensure any printed copies sent to journals are of good quality
- Use double spacing for the text
- Remember to add page numbers and possibly line numbers:
    - it will be easier for an editor to locate sections of your paper
- Read the journal's instructions to authors before submitting your paper:
    - formatting conventions vary between journals

Choosing a journal to publish your paper in

You should have an idea of a journal you would like to publish your paper in before you begin writing it. First, choose one that covers an appropriate subject area. If your work is quite specialised, it is useful to scan the tables of contents of journals to see if they publish articles in your field. You can also read the scope section of a journal, which is often found in the instructions to authors.

Second, you should aim to publish your paper in a prestigious journal. Your future progress (promotions and grants) may be affected by the journal your work appears in. A good way to rank the importance of journals is by 'impact factor'. This is a measure of how often a journal is cited and is defined as the 'average citations per article published'. Generally, the more prestigious a journal the higher its impact factor. You can find more about impact factors in the ISI Journal Citation Reports, accessed through the Web of Knowledge at *http://wok.mimas.ac.uk*. Other points to consider include the circulation of the journal and the frequency of publication (see 'Journal quality' in Chapter 3, page 41).

Impact factor top 10

Below is a list of the ten journals with the highest impact factors in 2003:

Journal	Impact Factor
1. Annual Review of Immunology	46.2
2. CA-A Cancer Journal for Clinicians	35.9
3. Annual Review of Biochemistry	31.6
4. Physiological Reviews	30.1

5.	Nature Genetics	29.6
6.	Cell	29.2
7.	New England Journal of Medicine	29.1
8.	Nature	28.0
9.	Nature Medicine	27.9
10.	Annual Review of Neuroscience	27.2

The journals in your field may not appear in a general list such as that provided above and so you may need to make a list that is specific to your subject area. However, it is quite likely that other scientists, or even some journals, have already done this for you and an Internet search for such compilations is worthwhile.

The peer-review process

When you do finally submit your paper to a journal, you are likely to have your work criticised, sometimes heavily, even if it is accepted. Do not be demoralised by this – even the most eminent scientists have to rework their papers. Responding to referees' comments can be stressful and requires experience. Advice from your senior co-authors and colleagues can be especially helpful at this point (see 'Peer-review' in Chapter 4, page 55).

Scientific fraud

Definition of fraud:

1. A deception deliberately practised in order to secure unfair or unlawful gain
2. A piece of trickery; a trick
3. a) One that defrauds; a cheat
   b) One who assumes a false pose; an impostor

Scientific fraud has become less rare and more publicised in recent years. The increase in the number of cases of misconduct may be partly due to escalating competition for grants and the pressure to publish work. Other reasons may include better monitoring and reporting mechanisms and an increased awareness of fraud in both scientists and non-scientists alike.

Scientific fraud can vary from the most extreme and obvious cases, for example, manufacturing data and plagiarism, to the more difficult borderline examples, such as omitting bad results. It is the obvious type of fraud that usually makes the headlines.

Dedicated organisations to tackle fraud

To deal with allegations of misconduct, a number of countries have set up dedicated organisations, such as the US Office of Research Integrity (*http://ori.dhhs.gov/*) and the Danish Committee on Scientific Dishonesty (*http://www.forsk.dk/eng/uvvu/index.htm*). The Office of Science and Technology (*http://www.ost.gov.uk/*) defines scientific misconduct as:

> 'fabrication, falsification, plagiarism, or other practices that seriously deviate from those that are commonly accepted within the scientific community for proposing, conducting, or reporting research. It does not include honest error or honest differences in interpretations or judgements of data'.

In order to promote good practice within science in the UK, the Committee on Publication Ethics (*http://www.publicationethics.org.uk/*) was set up in 1997 by British medical editors, including those of the *British Medical Journal*, *The Lancet* and *Gut*. The committee advises on cases of fraud, produces good practice guidelines and offers teaching and training within this area.

COPE and worldwide research organisations stress the importance of a number of practices in helping to prevent scientific misconduct:

- establish clear lines of responsibility in a research team
- conduct experiments in an open way
- keep good notes in your lab book
- archive data
- recognise the value of mentorship
- teach good practice to young scientists

There are varying degrees of plagiarism ranging from copying passages verbatim and passing them off as your own to rewriting someone else's ideas without acknowledging the author. Most institutions will provide guidance on plagiarism and even fraud.

Avoiding 'innocent' fraud

Some scientists deliberately set out to deceive, and may go to great lengths to publish their work and achieve prominence. However, researchers are often unaware that they are involved in fraudulent activities. Here are some potentially dubious practices:

- omitting 'non-fitting' data or the results of badly designed experiments without giving an explanation
- inappropriate treatment of data

- including citations that you have not read
- not citing other people's work that should be given credit
- omitting an author who has made an important contribution
- inclusion of an author who has not made a sufficient contribution (or who doesn't know their name is on the paper).

What to do if you suspect fraud?

Each employing institution should have its documented procedure for the investigation and management of complaints arising from fraud or misconduct, which should be consulted and followed explicitly.

Checklist

✓ Be organised
✓ Have the journals you want to submit to and the audience in your mind before you start
✓ Use a style that is clear, concise and consistent
✓ Avoid unnecessary jargon
✓ Prepare an accurate descriptive title; it is the most widely read part of your paper
✓ Use tables and figures
✓ Do not draw conclusions that are not supported by evidence present in the paper
✓ Reference thoroughly and make sure all work not your own is acknowledged as such
✓ Ask for feedback from your colleagues

Finding out more

Books

Booth, V. (1993) *Communicating in Science: Writing a scientific paper and speaking at scientific meetings, 2nd edition*, Cambridge University Press, Cambridge.
Council of Biology Editors (1994) *Scientific Style and Format: The CBE Manual for Authors, Editors and Publishers, 6th edition*, CBE, Colorado, USA.
Day, R. A. (1998) *How to Write and Publish a Scientific Paper, 5th edition*, Oryx Press, Westport, Connecticut.
Matthews, J., Bowen, J. and Matthews R. (2001) *Successful Scientific Writing: A step-by-step guide for biomedical scientists, 2nd edition*, Cambridge University Press, Cambridge.

Sides, C. H. (1992) *How to Write and Present Technical Information,* 3rd edition, Cambridge University Press, Cambridge.

 ## Website

Murray, J., *A guide to writing scientific papers.* Available at: *http://faculty.uca.edu/~jmurray/BIOL4425/lab/reports/guide/*

Chapter 7
Writing and Defending Your Thesis

by Dr Stan Venitt
Emeritus Reader in Cancer Studies, University of London

Introduction

As a PhD student you must complete and submit a thesis to obtain your degree. Having a clear idea of exactly what a thesis is and how it is structured before you begin writing up your work will make the process much easier. Your thesis is likely to be based on at least three years of work, so it is vital to learn how to organise your work and your time in order to make writing up a less daunting task.

This chapter provides advice and practical tips on writing your thesis – from getting started to submitting the finished work. You may also find it useful to have a look at Chapter 6: Writing a Paper, as there are some similarities between writing papers and theses.

Bear in mind the contribution that writing papers, presenting seminars and preparing annual progress reports can make to the overall preparation for your thesis. As well as being a useful way of taking stock they are a great opportunity to practise writing. Presenting seminars and writing papers are dealt with elsewhere in this book (Chapters 5 and 6 respectively) but there is some guidance on producing annual reports at the end of this chapter.

What is a thesis?

Every institution publishes regulations governing higher degrees, which include guidelines for the preparation and submission of theses. These regulations will probably be available on your institution's website. It is essential that you obtain a copy and prepare your thesis in light of these rules.

Although requirements for theses will vary from one institution to another, it may be useful to look at this example of a definition of a PhD thesis, taken from the University of London's code of practice, 2001:

'The thesis shall

(a) consist of the candidate's own account of his/her investigations and must indicate how they appear to him/her to advance the study of the subject;

(b) form a distinct contribution to the knowledge of the subject and afford evidence of originality by the discovery of new facts and/or by the exercise of independent critical power;

(c) be an integrated whole and present a coherent argument;

(d) give a critical assessment of the relevant literature, describe the method of research and its findings, and include a discussion on those findings, and indicate in what respects they appear to the candidate to advance the study of the subject;

(e) be written in English and the literary presentation shall be satisfactory;

(f) include a full bibliography and references;

(g) not exceed 100,000 words; a College may prescribe a lower number in certain subject areas, which shall be detailed in the relevant College regulations;

(h) be of a standard to merit publication in whole or in part or in a revised form (for example, as a monograph or as a number of articles in learned journals).'

So, your thesis is the culmination of your entire programme of study and is, in essence, a research report.

The purpose of a thesis

External examiners use the thesis to assess whether you have met the criteria needed to gain a higher degree. Your oral examination will be based on your thesis, which you will have to defend against the examiners' criticism and comments. Once you gain your PhD, prospective employers might wish to see your thesis, and its quality may affect your prospects. Moreover, the fact that you have completed a thesis tells prospective employers, in any field, that you have the determination, independence and stamina to complete a programme of study and research.

Researchers usually communicate with each other by publishing in scientific journals and by attending scientific conferences. However, other workers in your laboratory might want to consult your thesis to find out, for example, about a particular technique or piece of equipment you used. Scientists from other institutions working in the same field may also want to take a look at your thesis. They can do this by ordering microfilm versions from your institution's library. Advances in digital technology have meant that some theses are now available electronically – these 'e-theses' can be accessed easily from anywhere in the world and so potentially have a large audience (see 'Unpublished (grey) material' in Chapter 3, page 42.)

Have a plan

The thought of writing your thesis may seem a lot worse than actually doing it. A good way to start is to write a plan of your work. You can break down your research into manageable chunks and make a plan of all the sections and chapters. Once you have done this, you can keep breaking the sections down further, even to paragraph level. Now you only have to start writing a paragraph instead of a whole thesis.

As well as a plan, it can be useful to construct a timeline. On this you can list the date the thesis has to be finished and pencil in deadlines for different chapter and draft copies. It is a good idea to periodically update the timeline to change existing deadlines, if needed, and add any new ones. Setting clear goals for each week or even each day can be a good way to motivate yourself and manage your time. You will also need to decide which tasks should be completed first (see 'An overview of the planning phases' in Chapter 1, page 9).

Getting started

You don't have to wait until you have completed all of your laboratory work before you start writing your thesis. It is often best to write up particular experiments while they are still fresh in your mind, as otherwise you may forget important details. Before you begin writing a section it is essential to assemble as many of your results (tables, graphs, pictures and charts) and references as you can, as this will help you to structure the work. You don't have to write in a chronological order – it is often easier to start with a part you feel most comfortable with. If you have trouble getting started, try to write *something* – whether it is a draft of the contents page, references or just some notes.

Liaising with your supervisor

Your supervisor is likely to be your main source of support and guidance. You may wish to begin by discussing your plan with him or her and inviting suggestions for improving it. Your supervisor will have seen many more theses than you have and will be in a good position to advise you on content and structure.

Tips on your relationship with your supervisor

- Ensure that you maintain regular contact with your supervisor
- If you have difficulties with writing seek help early. Your institution may well have made arrangements for advice and instruction on writing and may provide formal writing courses
- Keep your supervisor informed of your progress (or lack of it). You may wish to provide him or her with completed chapters for review. Before doing this make sure that you have given due warning so that your supervisor can indicate when you can expect to receive his or her comments

- Do not expect your supervisor to write your thesis for you – his or her job is to give expert and timely criticism and advice, and for you to act upon it

Use of statistics

Always seek statistical advice before doing a set of experiments. Ideally, you should know at the planning stage which statistics you intend to apply to your data. If your institution has a department with statistical expertise it may be worthwhile establishing contact and seeking advice before embarking on what could be elaborate and expensive experiments. Make sure that you use the appropriate statistical tests and that your calculations are correct.

Use a software package only if you understand the statistical principles behind the tests you have used it for (see 'Statistics packages' later in this chapter). Make sure that you can answer simple questions about the tests you quote; if you can't, state that your statistics were done by a named statistician.

Know your audience

Your most important audience will be your external examiners. Usually, your thesis will be examined by one person who works at your institution and an examiner who works within your field but at an external institution. These people will be busy and will expect to see a well-structured and coherently argued thesis. Your thesis should also be aimed at researchers who want to refer to your work and repeat your experiments or methodology.

 Remember, you are the expert on the topic of your thesis and so others may need more background information than you think. Be explicit; don't make assumptions about your audience's level of expertise.

Background reading

As well as reading published scientific papers within your subject area, it is useful to have a look at the theses of previous graduates in your department (especially those who have gone on to do well). Doing this will not only help you understand the appropriate structure and style of presentation, but will also give you an idea of what the finished thesis will look like.

Organise your work

- Develop a filing system:
    - structure your computer file system in the same way as your thesis – create folders for each chapter and the references
    - have a similar paper-based filing system
- Back up your work regularly:
    - save computer files onto disks or another computer so that you always have two copies – take care to correctly date and label these
    - duplicate key parts of paper work such as laboratory books, diaries or manual records
- Design a system to distinguish between different drafts:
    - use different coloured paper or alternative fonts to help you quickly identify a draft

(See 'Organising your work' in Chapter 1, page 4.)

The structure of your thesis

Your thesis will be divided up into clear sections or chapters and the usual format is:

- Title page
- Abstract
- Acknowledgments
- Table of contents
- Introduction
- Materials and methods*
- Results
- Discussion
- Conclusion
- Experimental*
- References
- Appendices

(*The inclusion of these chapters and their position are discipline-dependent; if in doubt, check with your supervisor.)

Length

The maximum length of your thesis will be governed by the regulations of your institution and should not be exceeded. Make sure you are familiar with the rules before you start.

The order of writing

You do not have to write the chapters of your thesis in chronological order. It can be easier to start with the material that you are most familiar with. This way you also get into the flow of writing before you hit the trickier sections. Here is one suggestion for how you might order your writing:

1. Materials and methods or experimental
2. Results (including tables, charts and other graphical material)
3. Discussion
4. Conclusion
5. Introduction
6. References
7. Acknowledgments
8. Abstract

The reason for writing the introduction so near the end of the process is that you will have a clearer idea of the significance of the work after you have written up the results of your experiments and discussed them.

Writing your thesis

Each section or chapter of your thesis has specific requirements to fulfil. The purpose of each of the chapters has been summarised below.

The abstract

The abstract is a vital part of your thesis and it is best to do it last (but make sure you still spend enough time on it). The abstract is a summary of the work reported in your thesis. It should contain a description of the problem(s) addressed, the methodology used, the results and main conclusions. This is the part of your thesis that will be most widely read as it will be published in the ASLIB (Association of Special Libraries and Information Bureaux) Abstracts of Theses – it is important that the abstract can be understood on its own.

 Remember, do not exceed the word limit!

The introduction

The introduction will set the tone of your work and so it is important for it to be clear and concise. In this chapter you should review the background literature and state how your work contributes to the field of study and increases knowledge. You also need to establish the aims and objectives of your work, and briefly state the method of investigation and your main findings.

Materials and methods

In this chapter you should describe what you have done in sufficient detail to enable someone else to repeat your work. This is usually the easiest part to write and will provide practice for writing the more difficult sections.

Experimental

This section, generally found in chemistry theses, contains full details of the experimental procedures used to prepare compounds and the data associated with their identification. An experimental section comes after the conclusion and before the references.

Results

It is a good idea to restate your experiments briefly before presenting the relevant data. This will help your reader to understand the way the results are organised. Data should be digested and condensed; the use of tables and graphs is central to this section. Raw data should always be provided, but it can usually be assembled in an appendix (see 'Statistics packages and image file formats' later in this chapter).

Discussion

Here you will discuss what you have found from your experiments and the significance of this. It is pointless trying to describe the results to someone else until you have understood them yourself. This may require extensive manipulation of the data. For example, would data in a complicated table be more easily understood in the form of a graph? The use of spreadsheets and plotting software will allow you to explore your data in many different ways.

Data from replicate experiments should be reported separately. If you have not performed replicate experiments, you should be prepared to defend your decision.

It is useful to briefly restate your main results and to interpret them for the reader. Be specific and quote examples from your data. Don't just write, '*Table n shows that. . .*'. Extract an example and discuss it quantitatively. Avoid writing '*results of a typical experiment are shown. . .*' as this is likely to evoke questions about atypical experiments.

Bear in mind the following points when writing the discussion of your results:

- go from the particular to the general and discuss your results within the terms of the immediate study first
- place them in the context of work by others only when you have extracted all that you can from your own data
- give evidence of original thought; for example, you might propose an hypothesis or you might wish to demolish an hypothesis
- don't give the impression that you believe everything that you have read
- be critical and don't be afraid of revealing the deficiencies of your work
- suggest how the work would have been improved by what you learned from your mistakes.

You may find it easier to combine both the results and the discussion, as you will have several years of data. If you wait until the end of all the results, the reader may have some difficulty remembering what you are writing about. The results and discussion can then be subdivided according to subject matter.

Conclusion

This is where your main discoveries are summarised and emphasised. It is helpful to repeat the aims of the study in brief, so that your findings are linked to the original problems studied.

References

Whenever you mention information contained elsewhere, such as in a paper or book, you need to acknowledge the source. You should choose a recognised bibliographic style such as the Vancouver style used in *The Lancet* and the *British Medical Journal*, and stick to it. There are several bibliographic database software packages that are useful for assembling lists of references cited in a word processed document. These include Endnote and Reference Manager (see 'Managing your references' in Chapter 3, page 52). Make sure you check that the citations in the text match the references listed at the end of the thesis.

Tips on what to do if the writing stalls

- Talk through your ideas with peers
- Break down your work into manageable chunks
- Write your thoughts down as they come, then re-write
- Just make a start – with something easy
- Keep on writing, don't stop altogether just because you're stuck on one area

---

Look after yourself

- Take breaks:
    - intersperse long sessions of work with some short breaks (and try to vary the type of work you do)
- Exercise:
    - this can help to relieve stress and improve your productivity
- Eat properly:
    - don't forget to eat!
    - try to eat healthily and have regular meals
    - avoid drinking lots of coffee to try and stay awake

Statistics packages and image file formats

Statistics packages are computer programs that are used to manipulate and analyse data. Most packages use a number of statistical and graphical methods, which are designed for researchers from many different backgrounds. Two commonly used packages are SPSS (*http://www.spss.com/spss/*) and Stata (*http://www.stata. com/info/*), both of which and are straightforward to use. However, you should be careful about carrying out powerful statistical calculations without any knowledge of the background theory. Most packages have help menus that explain the concept of the analysis and how to enter your data, also many institutes run courses on statistics and available software for students.

You will probably need to store (in electronic format) a variety of charts and graphs showing your results. Images and graphics can be stored in a number of file formats, which have been developed over the years for specific applications and hardware. Below is a brief description of some of the more common file formats.

Graphics Interchange Format (.GIF)

GIF files are always compressed and so offer an efficient way to store large images. These colour-mapped files can have anywhere between 2 and 256 colours, which means that they are useful for limited colour images that are fairly simple with sharp definition, such as clipart and text. However, they should not be used for professional imaging and photographs. GIF files are used for online publishing.

Joint Photographic Experts Group (.JPG, .JPEG)

The JPEG file format uses a method of compression that reduces image file size by selectively reducing the amount of detail contained in the image. This means that, although the stored image will usually appear to be the same as the original, it will be missing some of its detail. Therefore JPEG files should not be used to store work in progress, but can be used for transmission of images by e-mail and for publishing images on the web, given their small size. The JPEG format is best for images with colour gradients such as photos, as it uses 24-bit colour.

Portable Network Graphics (.PNG)

This is a relatively new format that is the successor to GIF. It has better compression than GIF files, is good for the storage and transmission of most image types, and can handle 48-bit images. However, many older Internet browsers don't support PNG files.

Tag Image File Format (.TIFF, .TIF)

The TIF format is good for storing all types of images and is the standard for professional print publishing. Images are high in quality but often large in file size.

Windows Bitmap (.BMP)

This is the standard image file format used by Microsoft Windows. Most Windows Bitmap files are not compressed – their large size means that they are not efficient for the transmission and storage of images.

In summary GIF and JPEG files are a good choice if you want to e-mail files to your colleagues and supervisor. Where disk space and file size are not a problem, TIF files will provide you with high quality graphics.

Written style

You should aim to make your writing clear and unambiguous, so that what you say is not open to misinterpretation. Using short and simple words, phrases and sentences will help you to achieve this (although technical words will always be necessary). You may find it useful to construct a list of keywords that are central to your research, to use throughout your thesis – using several different words to mean the same thing may confuse your readers.

 Remember to be consistent with your style.

Formal language

The style of a thesis is quite formal. You should remember that scientific English is an international language and that the use of slang will not be understood by many. Looking at other successful theses will help you to understand and develop this style.

Spelling and grammar

Ensure that you use correct spellings and good grammar; it is helpful to have both a dictionary and a thesaurus handy when you are writing up your work. Microsoft Word contains an integral spell checker and thesaurus, and a number of separate spelling and grammar-checking programs are available for both PC and Macintosh computers. However, while these will be able to sort out simple grammatical errors, they will not be able to correct badly written prose. You should also avoid the use of jargon and buzz words, which can confuse a reader.

Active versus passive voice

Although the passive voice ('*The compound was weighed* . . .') has been used more commonly in the past, the active voice ('*I weighed the compound* . . .') is now encouraged by many scientists. This is because the active voice is usually simpler and requires fewer words. It also makes it clearer which parts of the work you were responsible for and which were done by other people.

For chemistry theses, however, particularly when you are writing experimental details, it is standard practice to use the passive voice. Decisions on active versus passive voice should be discussed carefully with your supervisor to ensure that an appropriate style, one that matches examiners' expectations, is adopted for the discipline.

Tips on written style

- Use adjectives and adverbs sparingly, for example, '*basically*', '*interesting(ly)*', '*obvious(ly)*', '*hopefully*' and '*typically*'
- Be precise, using words like '*situation*' and '*issue(s)*' may obscure rather than give clarity
- Some consider '*challenge*' and '*issue*' to be watered down versions of '*problem*'. For example, think about using '*solving problems*' instead of the less precise '*addressing issues*'

> - Minimise the use of buzz words such as *'paradigm'*, *'parameter'*, *'interaction'*, *'learning-curve'*, *'prioritise'*, *'leading-edge'* and *'interface'*
> - Every sentence in your thesis should contain something relevant
> - Critically examine every phrase for content, style and precision

 Remember, the published regulations for submitting your thesis will contain specific guidance on procedure and also on the presentation of the text. It is essential that you consult these rules *before* preparing your thesis.

The oral examination

After you have submitted your thesis you will be required to attend an oral examination (also known as a *viva voce* or *viva*). The purpose of this is to ensure you have a thorough knowledge of your research area and that you conducted the work yourself. The oral examination is an integral part of the degree examination and requires thorough preparation.

Preparation

There are a number of ways you can prepare for the oral examination:

- discuss the examination with your supervisor well before the event
- find out who will be examining you and what their interests and backgrounds are by researching and reading their papers
- ask your supervisor to arrange a mock *viva* with him or her and other scientists knowledgeable in your field of research
- practise presenting your research and discussing it with friends and colleagues
- keep up-to-date with developments within your research area – you will be required to know the most recent contributions to your field of study
- read and reread your thesis
- think about questions that may be asked and the answers to them – to help you do this think about some of the questions raised by your supervisor or asked of you during seminars
- if you can, try to attend someone else's mock oral examination before your own.

Some typical general questions that you should consider include:

- what did you enjoy most about your work?
- what would you do differently if you started all over again?

- what are the strongest or weakest parts of your work?
- what is original about your work?

What to expect

There will usually be two examiners present – one internal and one external to your institution. Your supervisor will not be an examiner, except in very exceptional circumstances, and will not normally be present at your *viva* unless you specifically request it. If your supervisor is present, he or she cannot intervene or assist you with questions. The examination will normally take between one and three hours. This may seem like a long time but once you get going you will probably find it goes much faster than you thought. You should expect your thesis to receive criticism – this is an accepted part of a *viva* examination. You should be robust in the defence of your thesis, but not to the point of giving offence. If, however, you are asked a question you cannot answer try not to prevaricate.

The outcome

You may be very lucky and not have to make any changes to your thesis following the oral examination. However, it is likely that you will have to make some amendments to the work in order to satisfy the examiners. In this case, the degree will be awarded but the modified thesis will usually have to be submitted within an agreed period of time.

Another possibility is that you will have to resubmit your thesis because it has some major weaknesses. You would normally have up to two years to do this. Rarer outcomes include being awarded a lower degree or no degree at all, but with the guidance of your supervisor this should be avoided. All these eventualities will be governed by your institution's regulations.

Tips on making modifications to your thesis

- Minor changes should not significantly change the structure:
  - 'white out' out mistakes and hand write in the correction over the top
  - alternatively, print out amended sentences/words and stick over the top
- Major changes may need more thought:
  - new pages and sections may have to be added
  - the thesis may need to be unbound to add in pages, or have new copies bound
  - ensure page numbering in the contents list is still correct

Checklist

✓ Practise writing throughout your PhD

✓ Set down your ideas and plan your thesis clearly in writing

✓ Don't leave writing up to the last minute

✓ Write the easiest parts first

✓ When writing your thesis, don't be satisfied with the first thing that comes into your head – rewriting and finessing is a very important process

✓ Give writing the time it deserves

✓ Be explicit, don't make assumptions about the reader's prior knowledge

✓ Keep full records of all your experiments

✓ Back up all your data

✓ Provide yourself with incentives

✓ Choose your style for citations and references and use it from the start

✓ Arrange a mock *viva*

✓ Keep up-to-date with the literature in your research area

✓ Anticipate and prepare for questions in your viva

A quick guide to writing annual progress reports

Students are often required to write annual progress reports on their PhD work. These reports serve two functions:

1. To train you in the clear presentation of scientific information
2. To keep your institution informed of your progress

Such reports are an important part of your training and will help make the writing of your thesis easier; they are usually appraised and should be prepared carefully and written in good English.

General points

- Reports should be easily understood as discrete entities
- The introductory paragraph should outline clearly and succinctly the project and its relevance to the advancement of knowledge in your chosen field
- Reports should be prepared by you, but supervisors should give guidance on the clear presentation of scientific information and on literary style
- Supervisors should check that all statistical analysis is accurate
- Experiments at a preliminary stage (which have not been repeated or proved statistically) should be included but interpreted with caution
- You should discuss the structure of your report with your supervisor before writing

 Remember, there may be presentation and submission requirements for your report and you should check any guidance in advance.

As an example, here is a summary of The Institute of Cancer Research's requirement for first, second and final year reports. Other institutions may not require a final year report, and some may ask for a poster presentation in the second year instead of a report.

First year report

- Provide an abstract of about 200 words, followed by a comprehensive report of between 3000 and 5000 words
- Write an introduction stating the nature of the problem under investigation
- Give a plan of the work to be done, including a survey of the proposed experimental methods
- Finally, present a summary and discussion of the results obtained so far, together with comments on the future development of your project
- Include key references in the report
- Use diagrams to illustrate particular points, but don't use them to excess and always give them adequate explanation in the text
- Do not exceed the stated word limit

Second year report

- Start with a concise outline of the project – this is essential as the assessor of an intermediate report may not necessarily be the same as for the first year report
- Provide an introduction, a summary of total progress to date and a statement of what remains to be done
- Include adequate presentation of data (e.g. figures and tables) to allow a proper assessment of the work
- Attach a copy of your First Year Report to minimise the need for repetition
- Include key references
- You should also prepare a plan, in conjunction with your supervisors, of how your work will progress through the third and final year of study to completion
- In the plan, highlight any actual or potential problems and the means of resolving them
- Include a realistic estimation of the completion date
- This plan should give the following information:
  - content of the thesis
  - tasks outstanding and time period during which these will be completed

Final year report

- Prepare an outline thesis including a 300-word thesis description (which will be needed when submitting the examination entry if this has not already been done)
- Give a realistic estimate of the completion date
- Provide details of the contents of each chapter in brief, including details of the experiments carried out, and highlighting any further experiments which will be needed to complete the project

Finding out more

Books

Bolker, J. (1998) *Writing Your Dissertation in Fifteen Minutes a Day: A guide to starting, revising, and finishing your doctoral thesis*, Owl Books, New York.

Cryer, P. (2000) *The Research Student's Guide to Success*, 2nd edition, Open University Press, Buckingham.

Davis, G. and Parker, C. (1997) *Writing a Doctoral Dissertation: A systematic approach*, 2nd edition, Barron's Educational Series, Hauppauge, New York.

Day, R. A. (1998) *How to Write and Publish a Scientific Paper*, 5th edition, Oryx Press, Westport, Connecticut.

Dunleavey, P. (2003) *Authoring a PhD Thesis: How to plan, draft, write and finish a doctoral dissertation (Palgrave study guides)*, Palgrave Macmillan, Basingstoke.

Kirkman, J. (1992) *Good Style: Writing for science and technology*, E & FN Spon, London.

Murray, R. (2002) *How to Write a Thesis*, Open University Press, Buckingham.

Websites

Levine, S. J. (1999) *Writing and Presenting your Thesis or Dissertation*. Available at: *http://rses.anu.edu.au/gfd/Gfd_user_links/andrew.kiss.directory/thesis_writing/ thesis_guide.html*

Information Services and Systems, King's College London website. Scientific writing and publishing: writing up a project, thesis or dissertation. Available at: *http://www.kcl.ac.uk/depsta/iss/schools/writing/writingup.html*

Index

Page numbers in italics indicate figures or tables.

Index

Index

Printed in the United States
142414LV00007B/3/P

9 780470 094853